Florian Ion PETRESCU &
Relly Victoria PETRESCU

MECHANICAL ENGINEERING
DESIGN I

Germany 2012

Scientific reviewer:

Dr. Veturia CHIROIU

Honorific member of
Technical Sciences Academy of Romania (ASTR)
PhD supervisor in Mechanical Engineering

Copyright

Title book: Mechanical Engineering Design I
Authors book: Florian Ion Petrescu & Relly Victoria Petrescu

© 2011-2012, Florian Ion PETRESCU
petrescuflorian@yahoo.com

ALL RIGHTS RESERVED. This book contains material protected under International and Federal Copyright Laws and Treaties. Any unauthorized reprint or use of this material is prohibited. No part of this book may be reproduced or transmitted in any form or by any means, electronic or mechanical, including photocopying, recording, or by any information storage and retrieval system without express written permission from the authors / publisher.

**Manufactured and published by:
Books on Demand GmbH, Norderstedt**

ISBN 978-3-8482-3014-3

WELCOME

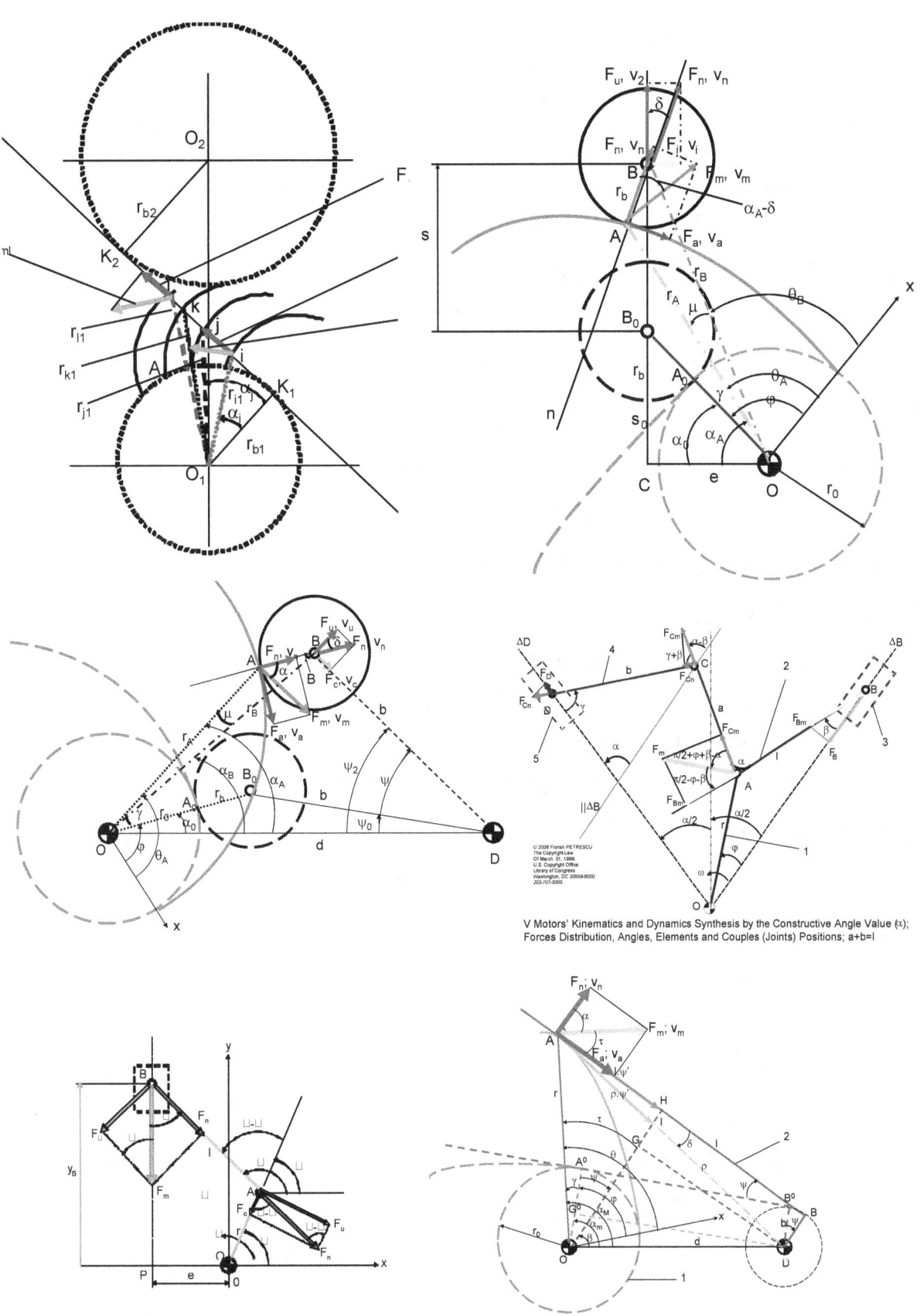

V Motors' Kinematics and Dynamics Synthesis by the Constructive Angle Value (α); Forces Distribution, Angles, Elements and Couples (Joints) Positions; a+b=l

You are welcome to read the full book! The authors.

CONTENT

Welcome 003
Content 004
Cap 01 FORCES AND EFFICIENCY OF GEARING 005
Cap 02 GEAR DYNAMIC SYNTHESIS 012
Cap 03 PRESENTING A "DYNAMIC ORIGINAL MODEL" USED TO STUDY TOOTHED GEARING WITH PARALLEL AXES 019
Cap 04 PLANETARY TRAINS EFFICIENCY 026
Cap 05 CAM GEARS EFFICIENCY 032
Cap 06 CONTRIBUTIONS AT THE DYNAMIC OF CAMS 039
Cap 07 CAM GEARS DYNAMICS ILLUSTRATED IN THE CLASSIC DISTRIBUTION 046
Cap 08 CAM GEARS DYNAMICS TO THE MODULE B (WITH TRANSLATED FOLLOWER WITH ROLL) 056
Cap 09 CINEMATICS OF THE 3R DYAD 066
Cap 10 KINEMATICS OF THE PLANAR QUADRILATERAL MECHANISM 073
Cap 11 DETERMINING THE MECHANICAL EFFICIENCY OF OTTO ENGINE'S MECHANISM 079
Cap 12 OTTO ENGINE DYNAMICS 086
Cap 13 AN ORIGINAL INTERNAL COMBUSTION ENGINE 093
Cap 14 V ENGINE DESIGN 102

CHAPTER I

FORCES AND EFFICIENCY OF GEARING

ABSTRACT: *The chapter presents an original method to determine the efficiency of the gear, the forces of the gearing, the velocities and the powers. The originality of this method relies on the eliminated friction modulus. The chapter is analyzing the influence of a few parameters concerning gear efficiency. These parameters are: z_1 - the number of teeth for the primary wheel of gear; z_2 - the number of teeth of the secondary wheel of gear; α_0 - the normal pressure angle on the divided circle; β - the inclination angle. With the relations presented in this paper, it can synthesize the gear's mechanisms. Today, the gears are present everywhere, in the mechanical's world (In vehicle's industries, in electronics and electro-technique equipments, in energetically industries, etc...). Optimizing this mechanism (the gears mechanism), we can improve the functionality of the transmissions with gears.*

Keywords: Efficiency, forces, powers, velocities, gear, constructive parameters, teeth, outside circle, wheel.

1. INTRODUCTION

In this chapter the authors present an original method for calculating the efficiency of the gear, the forces of the gearing, the velocities and the powers.

The originality consists in the way of determination of the gear efficiency because it hasn't used the friction forces of couple (this new way eliminates the classical method).

It eliminates the necessity of determining the friction coefficients by different experimental methods as well.

The efficiency determinates by the new method is the same like the classical efficiency, namely the mechanical efficiency of the gear.

Precisely one determines the dynamics efficiency, but at the gears transmissions, the dynamics efficiency is the same like the mechanical efficiency; this is a greater advantage of the gears transmissions.

2. DETERMINING THE MOMENTARY DYNAMIC (MECHANICAL) EFFICIENCY, THE FORCES OF THE GEARING, AND THE VELOCITIES

The calculating relations [2, 3], are the next (1-21), (see the fig. 1):

$$\begin{cases} F_\tau = F_m \cdot \cos\alpha_1 \\ F_\psi = F_m \cdot \sin\alpha_1 \\ v_2 = v_1 \cdot \cos\alpha_1 \\ v_{12} = v_1 \cdot \sin\alpha_1 \\ \overline{F}_m = \overline{F}_\tau + \overline{F}_\psi \\ \overline{v}_1 = \overline{v}_2 + \overline{v}_{12} \end{cases} \quad (1)$$

with: F_m - the motive force (the driving force);

F_τ - the transmitted force (the useful force);

F_ψ - the slide force (the lost force);

v_1 - the velocity of element 1, or the speed of wheel 1 (the driving wheel);

v_2 - the velocity of element 2, or the speed of wheel 2 (the driven wheel);

v_{12} - the relative speed of the wheel 1 in relation with the wheel 2 (this is a sliding speed).

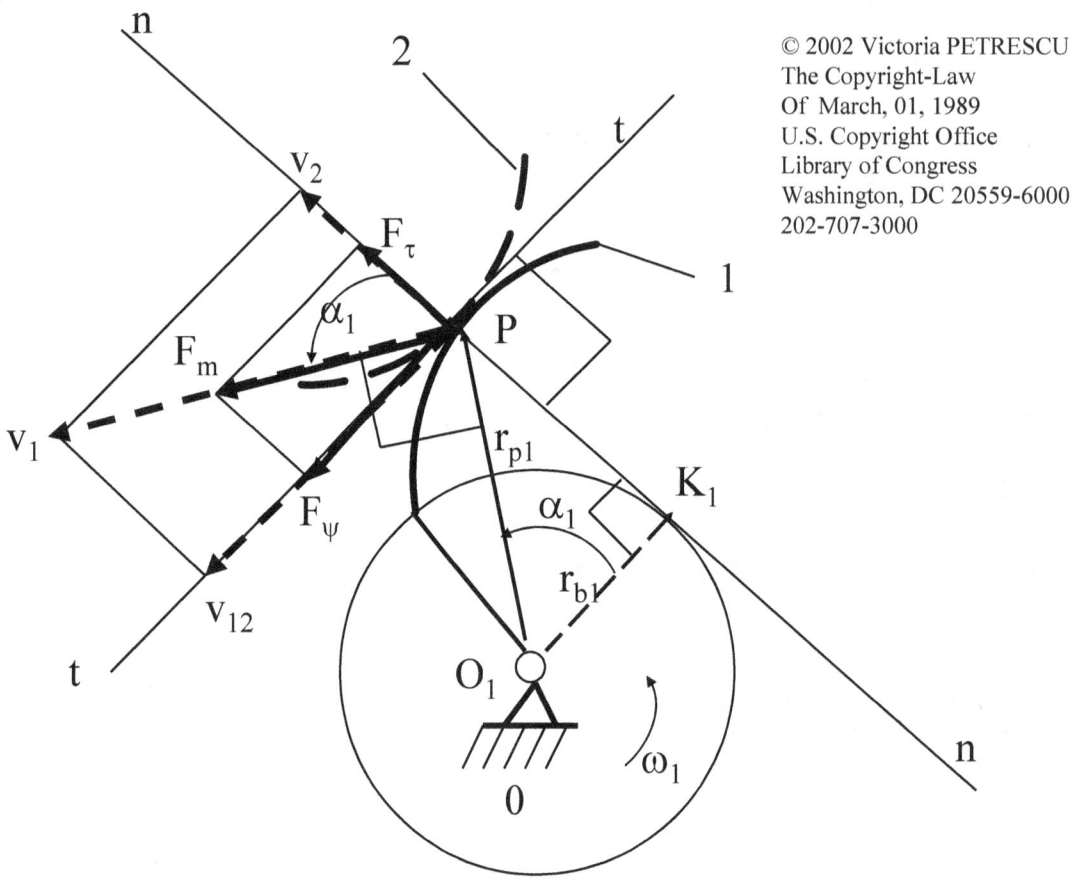

Fig. 1. *The forces and the velocities of the gearing*

The consumed power (in this case the driving power):
$$P_c \equiv P_m = F_m \cdot v_1 \qquad (2)$$
The useful power (the transmitted power from the profile 1 to the profile 2) will be written:
$$P_u \equiv P_\tau = F_\tau \cdot v_2 = F_m \cdot v_1 \cdot \cos^2 \alpha_1 \qquad (3)$$
The lost power will be written:
$$P_\psi = F_\psi \cdot v_{12} = F_m \cdot v_1 \cdot \sin^2 \alpha_1 \qquad (4)$$
The momentary efficiency of couple will be calculated directly with the next relation:

$$\begin{cases} \eta_i = \dfrac{P_u}{P_c} \equiv \dfrac{P_\tau}{P_m} = \dfrac{F_m \cdot v_1 \cdot \cos^2 \alpha_1}{F_m \cdot v_1} \\ \eta_i = \cos^2 \alpha_1 \end{cases} \qquad (5)$$

The momentary losing coefficient [1], will be written:

$$\begin{cases} \psi_i = \dfrac{P_\psi}{P_m} = \dfrac{F_m \cdot v_1 \cdot \sin^2 \alpha_1}{F_m \cdot v_1} = \sin^2 \alpha_1 \\ \eta_i + \psi_i = \cos^2 \alpha_1 + \sin^2 \alpha_1 = 1 \end{cases} \qquad (6)$$

It can easily see that the sum of the momentary efficiency and the momentary losing coefficient is 1:

Now one can determine the geometrical elements of gear. These elements will be used in determining the couple efficiency, η.

3. THE GEOMETRICAL ELEMENTS OF THE GEAR

We can determine the next geometrical elements of the external gear, [2,3], (for the right teeth, β=0):

The radius of the basic circle of wheel 1 (of the driving wheel), (7):

$$r_{b1} = \frac{1}{2} \cdot m \cdot z_1 \cdot \cos \alpha_0 \qquad (7)$$

The radius of the outside circle of wheel 1 (8):

$$r_{a1} = \frac{1}{2} \cdot (m \cdot z_1 + 2 \cdot m) = \frac{m}{2} \cdot (z_1 + 2) \qquad (8)$$

It determines now the maximum pressure angle of the gear (9):

$$\cos \alpha_{1M} = \frac{r_{b1}}{r_{a1}} = \frac{\frac{1}{2} \cdot m \cdot z_1 \cdot \cos \alpha_0}{\frac{1}{2} \cdot m \cdot (z_1 + 2)} = \frac{z_1 \cdot \cos \alpha_0}{z_1 + 2} \qquad (9)$$

And now one determines the same parameters for the wheel 2, the radius of basic circle (10) and the radius of the outside circle (11) for the wheel 2:

$$r_{b2} = \frac{1}{2} \cdot m \cdot z_2 \cdot \cos \alpha_0 \qquad (10)$$

$$r_{a2} = \frac{m}{2} \cdot (z_2 + 2) \qquad (11)$$

Now it can determine the minimum pressure angle of the external gear (12, 13):

$$\begin{cases} tg\alpha_{1m} = \dfrac{N}{r_{b1}} \\ N = (r_{b1} + r_{b2}) \cdot tg\alpha_0 - \sqrt{r_{a2}^2 - r_{b2}^2} = \\ \quad = \dfrac{1}{2} \cdot m \cdot (z_1 + z_2) \cdot \sin\alpha_0 - \dfrac{m}{2} \cdot \sqrt{(z_2 + 2)^2 - z_2^2 \cdot \cos^2 \alpha_0} = \\ \quad = \dfrac{m}{2} \cdot [(z_1 + z_2) \cdot \sin\alpha_0 - \sqrt{z_2^2 \cdot \sin^2 \alpha_0 + 4 \cdot z_2 + 4}] \end{cases} \qquad (12)$$

$$tg\alpha_{1m} = [(z_1 + z_2) \cdot \sin\alpha_0 - \sqrt{z_2^2 \cdot \sin^2 \alpha_0 + 4 \cdot z_2 + 4}]/(z_1 \cdot \cos\alpha_0) \qquad (13)$$

Now we can determine, for the external gear, the minimum (13) and the maximum (9) pressure angle for the right teeth. For the external gear with bended teeth ($\beta \neq 0$) it uses the relations (14, 15 and 16):

$$tg\alpha_t = \frac{tg\alpha_0}{\cos\beta} \qquad (14)$$

$$tg\alpha_{1m} = [(z_1 + z_2) \cdot \frac{\sin\alpha_t}{\cos\beta} - \sqrt{z_2^2 \cdot \frac{\sin^2\alpha_t}{\cos^2\beta} + 4 \cdot \frac{z_2}{\cos\beta} + 4}] \cdot \frac{\cos\beta}{z_1 \cdot \cos\alpha_t} \qquad (15)$$

$$\cos\alpha_{1M} = \frac{\frac{z_1 \cdot \cos\alpha_t}{\cos\beta}}{\frac{z_1}{\cos\beta} + 2} \qquad (16)$$

For the internal gear with bended teeth ($\beta \neq 0$) it uses the relations (14 with 17, 18-A, or with 19, 20-B):

A. When the driving wheel 1, has external teeth:

$$tg\alpha_{1m} = [(z_1 - z_2) \cdot \frac{\sin\alpha_t}{\cos\beta} + \sqrt{z_2^2 \cdot \frac{\sin^2\alpha_t}{\cos^2\beta} - 4 \cdot \frac{z_2}{\cos\beta} + 4}] \cdot \frac{\cos\beta}{z_1 \cdot \cos\alpha_t} \qquad (17)$$

$$\cos\alpha_{1M} = \frac{\frac{z_1 \cdot \cos\alpha_t}{\cos\beta}}{\frac{z_1}{\cos\beta} + 2} \qquad (18)$$

B. When the driving wheel 1, have internal teeth:

$$tg\alpha_{1M} = [(z_1 - z_2) \cdot \frac{\sin\alpha_t}{\cos\beta} + \sqrt{z_2^2 \cdot \frac{\sin^2\alpha_t}{\cos^2\beta} + 4 \cdot \frac{z_2}{\cos\beta} + 4}] \cdot \frac{\cos\beta}{z_1 \cdot \cos\alpha_t} \qquad (19)$$

$$\cos\alpha_{1m} = \frac{\frac{z_1 \cdot \cos\alpha_t}{\cos\beta}}{\frac{z_1}{\cos\beta} - 2} \qquad (20)$$

4. DETERMINING THE EFFICIENCY

The efficiency of the gear will be calculated through the integration of momentary efficiency on all sections of gearing movement, namely from the minimum pressure angle to the maximum pressure angle, the relation (21), [2, 3].

$$\eta = \frac{1}{\Delta\alpha} \cdot \int_{\alpha_m}^{\alpha_M} \eta_i \cdot d\alpha = \frac{1}{\Delta\alpha} \int_{\alpha_m}^{\alpha_M} \cos^2\alpha \cdot d\alpha = \frac{1}{2 \cdot \Delta\alpha} \cdot [\frac{1}{2} \cdot \sin(2 \cdot \alpha) + \alpha]_{\alpha_m}^{\alpha_M} =$$

$$= \frac{1}{2 \cdot \Delta\alpha} [\frac{\sin(2\alpha_M) - \sin(2\alpha_m)}{2} + \Delta\alpha] = \frac{\sin(2 \cdot \alpha_M) - \sin(2 \cdot \alpha_m)}{4 \cdot (\alpha_M - \alpha_m)} + 0.5 \qquad (21)$$

Table 1. Determining the efficiency of the gear's right teeth for $i_{12effective} = -4$

$i_{12effective} = -4$	right teeth					
$z_1 = 8$	$z_2 = 32$			$z_1 = 30$	$z_2 = 120$	
$\alpha_0 = 20°$?	$\alpha_0 = 29°$	$\alpha_0 = 35°$		$\alpha_0 = 15°$	$\alpha_0 = 20°$	$\alpha_0 = 30°$
$\alpha_m = -16.22°$?	$\alpha_m = 0.7159°$	$\alpha_m =$		$\alpha_m = 1.5066°$	$\alpha_m = 9.5367°$	α_m
$\alpha_M = 41.2574°$	$\alpha_M = 45.5974$	$\alpha_M = 49.056$		$\alpha_M = 25.1018°$	$\alpha_M = 28.241$	$\alpha_M = 35.718$
	$\eta = 0.8111$	$\eta = 0.7308$		$\eta = 0.9345$	$\eta = 0.8882$	$\eta = 0.7566$
$z_1 = 10$	$z_2 = 40$			$z_1 = 90$	$z_2 = 360$	
$\alpha_0 = 20°$?	$\alpha_0 = 26°$	$\alpha_0 = 30°$		$\alpha_0 = 8°$?	$\alpha_0 = 9°$	$\alpha_0 = 20°$
$\alpha_m = -9.89°$?	$\alpha_m = 1.3077°$	$\alpha_m =$		$\alpha_m = -0.1638°$	$\alpha_m = 1.5838°$	α_m
$\alpha_M = 38.4568°$	$\alpha_M = 41.4966$	$\alpha_M = 43.806$		$\alpha_M = 14.3637°$	$\alpha_M = 14.935$	$\alpha_M = 23.181$
	$\eta = 0.8375$	$\eta = 0.7882$			$\eta = 0.9750$	$\eta = 0.8839$
$z_1 = 18$	$z_2 = 72$					
$\alpha_0 = 19°$	$\alpha_0 = 20°$	$\alpha_0 = 30°$				
$\alpha_m = 0.9860°$	$\alpha_m = 2.7358°$	α_m				
$\alpha_M = 31.6830°$	$\alpha_M = 32.2505$	$\alpha_M = 38.792$				
$\eta = 0.90105$	$\eta = 0.8918$	$\eta = 0.7660$				

5. THE CALCULATED EFFICIENCY OF THE GEAR

We shall now see four tables with the calculated efficiency depending on the input parameters and once we proceed with the results we will draw some conclusions.

The input parameters are:

z_1 = the number of teeth for the driving wheel 1;

z_2 = the number of teeth for the driven wheel 2, or the ratio of transmission, i ($i_{12} = -z_2/z_1$);

α_0 = the pressure angle normal on the divided circle;

β = the bend angle.

Table 2. Determining the efficiency of the gear's right teeth for $i_{12effective} = -2$

$i_{12effective} = -2$	right teeth					
$z_1 = 8$	$z_2 = 16$			$z_1 = 18$	$z_2 = 36$	
$\alpha_0 = 20°$?	$\alpha_0 = 28°$	$\alpha_0 = 35°$		$\alpha_0 = 18°$	$\alpha_0 = 20°$	$\alpha_0 = 30°$
$\alpha_m = -12.65°$?	$\alpha_m = 0.9149°$	α_m		$\alpha_m = 0.6756°$	$\alpha_m = 3.9233°$	$\alpha_m = 18.6935°$
$\alpha_M = 41.2574°$	$\alpha_M = 45.0606$	$\alpha_M = 49.0559$		$\alpha_M = 31.1351°$	$\alpha_M = 32.250$	$\alpha_M = 38.7922°$
	$\eta = 0.8141$	$\eta = 0.7236$		$\eta = 0.9052$	$\eta = 0.8874$	$\eta = 0.7633$
$z_1 = 10$	$z_2 = 20$			$z_1 = 90$	$z_2 = 180$	
$\alpha_0 = 20°$?	$\alpha_0 = 25°$	$\alpha_0 = 30°$		$\alpha_0 = 8°$	$\alpha_0 = 20°$	$\alpha_0 = 30°$
$\alpha_m = -7.13°$?	$\alpha_m = 1.3330°$	$\alpha_m = 9.4106°$		$\alpha_m = 0.5227°$	α_m	$\alpha_m = 27.7825°$
$\alpha_M = 38.4568°$	$\alpha_M = 40.9522$	$\alpha_M = 43.8060$		$\alpha_M = 14.3637°$	$\alpha_M = 23.181$	$\alpha_M = 32.0917°$
	$\eta = 0.8411$	$\eta = 0.7817$		$\eta = 0.9785$	$\eta = 0.8836$	$\eta = 0.7507$

Table 3. Determining the efficiency of the gear's right teeth for $i_{12effective} = -6$

$i_{12effective} = -6$	right teeth				
$z_1 = 8$	$z_2 = 48$		$z_1 = 18$	$z_2 = 108$	
$\alpha_0 = 20°$?	$\alpha_0 = 30°$	$\alpha_0 = 35°$	$\alpha_0 = 19°$	$\alpha_0 = 20°$	$\alpha_0 = 30°$
$\alpha_m = -17.86°$?	$\alpha_m = 1.7784°$	$\alpha_m = 10.660°$	$\alpha_m = 0.4294°$	$\alpha_m = 2.2449°$	α_m
$\alpha_M = 41.2574°$	$\alpha_M = 46.1462°$	$\alpha_M = 49.0559$	$\alpha_M = 31.6830°$	$\alpha_M = 32.2505°$	$\alpha_M = 38.792$
	$\eta = 0.8026$	$\eta = 0.7337$	$\eta = 0.9028$	$\eta = 0.8935$	$\eta = 0.7670$
$z_1 = 10$	$z_2 = 60$		$z_1 = 90$	$z_2 = 540$	
$\alpha_0 = 20°$?	$\alpha_0 = 26°$	$\alpha_0 = 30°$	$\alpha_0 = 9°$	$\alpha_0 = 20°$	$\alpha_0 = 30°$
$\alpha_m = -11.12°$?	$\alpha_m = 0.6054°$	$\alpha_m =$	$\alpha_m = 1.3645°$	$\alpha_m = 16.4763°$	α_m
$\alpha_M = 38.4568°$	$\alpha_M = 41.4966°$	$\alpha_M = 43.8060$	$\alpha_M = 14.9354°$	$\alpha_M = 23.1812°$	$\alpha_M = 32.091$
	$\eta = 0.8403$	$\eta = 0.7908$	$\eta = 0.9754$	$\eta = 0.8841$	$\eta = 0.7509$

We begin with the right teeth (the toothed gear), with i=-4, once for z_1 we shall take successively different values, rising from 8 teeth. It can see that for 8 teeth of the driving wheel the standard pressure angle, $\alpha_0 = 20°$, is so small to be used (it obtains a minimum pressure angle, α_m, negative and this fact is not admitted!). In the second table we shall diminish (in module) the value for the ratio of transmission, i, from 4 to 2. It will see how for a lower value of the number of teeth of the wheel 1, the standard pressure angle ($\alpha_0 = 20°$) is to small and it will be necessary to increase it to a minimum value. For example, if $z_1 = 8$, the necessary minimum value is $\alpha_0 = 29°$ for i=-4 (see the table 1) and $\alpha_0 = 28°$ for i=-2 (see the table 2). If $z_1 = 10$, the necessary minimum pressure angle is $\alpha_0 = 26°$ for i=-4 (see the table 1) and $\alpha_0 = 25°$ for i=-2 (see the table 2).

When the number of teeth of the wheel 1 increases, it can decrease the normal pressure angle, α_0. One shall see that for $z_1 = 90$ it can take less for the normal pressure angle (for the pressure angle of reference), $\alpha_0 = 8°$. In the table 3 it increases the module of i, value (for the ratio of transmission), from 2 to 6.

In the table 4, the teeth are bended ($\beta \neq 0$). The module i, take now the value 2.

Table 4. The determination of the gear's parameters in bend teeth for i=-4

$i_{12effective} = -4$	bend teeth	Table 4			
	$\beta = 15°$				
$z_1 = 8$	$z_2 = 32$		$z_1 = 30$	$z_2 = 120$	
$\alpha_0 = 20°$?	$\alpha_0 = 30°$	$\alpha_0 = 35°$	$\alpha_0 = 15°$	$\alpha_0 = 20°$	$\alpha_0 = 30°$
$\alpha_m = -16.836°$	$\alpha_m = 1.1265°$	$\alpha_m = 9.4455°$	$\alpha_m = 1.0269°$	$\alpha_m = 8.8602°$	$\alpha_m = 22.1550°$
$\alpha_M = 41.0834°$	$\alpha_M = 46.2592°$	$\alpha_M = 49.2953°$	$\alpha_M = 25.1344°$	$\alpha_M = 28.4591°$	$\alpha_M = 36.2518°$
	$\eta = 0.8046$	$\eta = 0.7390$	$\eta = 0.9357$	$\eta = 0.8899$	$\eta = 0.7593$
$z_1 = 18$	$z_2 = 72$		$z_1 = 90$	$z_2 = 360$	
$\alpha_0 = 19°$	$\alpha_0 = 20°$	$\alpha_0 = 30°$	$\alpha_0 = 9°$	$\alpha_0 = 20°$	$\alpha_0 = 30°$
$\alpha_m = 0.32715°$	$\alpha_m = 2.0283°$	$\alpha_m = 17.1840°$	$\alpha_m = 1.3187°$	$\alpha_m = 15.8944°$	$\alpha_m = 26.9403°$
$\alpha_M = 31.7180°$	$\alpha_M = 32.3202°$	$\alpha_M = 39.1803°$	$\alpha_M = 14.9648°$	$\alpha_M = 23.6366°$	$\alpha_M = 32.8262°$
$\eta = 0.9029$	$\eta = 0.8938$	$\eta = 0.7702$	$\eta = 0.9754$	$\eta = 0.8845$	$\eta = 0.7513$

6. CONCLUSIONS

The efficiency (of the gear) increases when the number of teeth for the driving wheel 1, z_1, increases too and when the pressure angle, α_0, diminishes; z_2 or i_{12} are not so much influence about the efficiency value;

It can easily see that for the value $\alpha_0=20^0$, the efficiency takes roughly the value $\eta\approx0.89$ for any values of the others parameters (this justifies the choice of this value, $\alpha_0=20^0$, for the standard pressure angle of reference).

The better efficiency may be obtained only for a $\alpha_0 \neq 20^0$.

But the pressure angle of reference, α_0, can be decreased the same time the number of teeth for the driving wheel 1, z_1, increases, to increase the gears' efficiency;

Contrary, when we desire to create a gear with a low z_1 (for a less gauge), it will be necessary to increase the α_0 value, for maintaining a positive value for α_m (in this case the gear efficiency will be diminished);

When β increases, the efficiency, η, increases too, but the growth is insignificant.

The module of the gear, m, has not any influence on the gear's efficiency value.

When α_0 is diminished it can take a higher normal module, for increasing the addendum of teeth, but the increase of the module m at the same time with the increase of the z_1 can lead to a greater gauge.

The gears' efficiency, η, is really a function of α_0 and z_1: $\eta=f(\alpha_0,z_1)$; α_m and α_M are just the intermediate parameters.

For a good projection of the gear, it's necessary a z_1 and a z_2 greater than 30-60; but this condition may increase the gauge of mechanism.

In this chapter it determines precisely, the dynamics-efficiency, but at the gears transmissions, the dynamics efficiency is the same like the mechanical efficiency; this is a greater advantage of the gears transmissions. This advantage, specifically of the gear's mechanisms, may be found at the cam mechanisms with plate followers as well.

REFERENCES

[1] Pelecudi, Chr., ş.a., *Mecanisme*. E.D.P., Bucureşti, 1985.
[2] Petrescu, V., Petrescu, I., *Randamentul cuplei superioare de la angrenajele cu roţi dinţate cu axe fixe*, In: The Proceedings of 7th National Symposium PRASIC, Braşov, vol. I, pp. 333-338, 2002.
[3] Petrescu, R., Petrescu, F., *The gear synthesis with the best efficiency*, In: The Proceedings of ESFA'03, Bucharest, vol. 2, pp. 63-70, 2003.

CHAPTER II

GEAR DYNAMIC SYNTHESIS

Abstract: *In this chapter one succinctly presents an original method to obtain the efficiency of geared transmissions in function of the contact ratio of the gearing. With the presented relations it can make the dynamic synthesis of geared transmissions, having in view increasing the efficiency of gearing mechanisms.*
Keywords: *gear efficiency, gear dynamics, contact ratio, gear synthesis*

1 Introduction

In this chapter one presents shortly an original method to obtain the efficiency of the geared transmissions in function of the contact ratio. With the presented relations it can make the dynamic synthesis of the geared transmissions having in view increasing the efficiency of gearing mechanisms in work.

2 Determining of gearing efficiency in function of the contact ratio

We calculate the efficiency of a geared transmission, having in view the fact that at one moment there are several couples of teeth in contact, and not just one.

The start model has got four pairs of teeth in contact (4 couples) concomitantly.

The first couple of teeth in contact has the contact point i, defined by the ray r_{i1}, and the pressure angle α_{i1}; the forces which act at this point are: the motor force F_{mi}, perpendicular to the position vector r_{i1} at i and the force transmitted from the wheel 1 to the wheel 2 through the point i, $F_{\tau i}$, parallel to the path of action and with the sense from the wheel 1 to the wheel 2, the transmitted force being practically the projection of the motor force on the path of action; the defined velocities are similar to the forces (having in view the original kinematics, or the precise kinematics adopted); the same parameters will be defined for the next three points of contact, j, k, l (Fig. 1).

For starting we write the relations between the velocities (1):

$$\begin{aligned} v_{\tau i} &= v_{mi} \cdot \cos\alpha_i = r_i \cdot \omega_1 \cdot \cos\alpha_i = r_{b1} \cdot \omega_1 \\ v_{\tau j} &= v_{mj} \cdot \cos\alpha_j = r_j \cdot \omega_1 \cdot \cos\alpha_j = r_{b1} \cdot \omega_1 \\ v_{\tau k} &= v_{mk} \cdot \cos\alpha_k = r_k \cdot \omega_1 \cdot \cos\alpha_k = r_{b1} \cdot \omega_1 \\ v_{\tau l} &= v_{ml} \cdot \cos\alpha_l = r_l \cdot \omega_1 \cdot \cos\alpha_l = r_{b1} \cdot \omega_1 \end{aligned} \quad (1)$$

From relations (1), one obtains the equality of the tangential velocities (2), and makes explicit the motor velocities (3):

$$v_{\tau i} = v_{\tau j} = v_{\tau k} = v_{\tau l} = r_{b1} \cdot \omega_1 \quad (2)$$

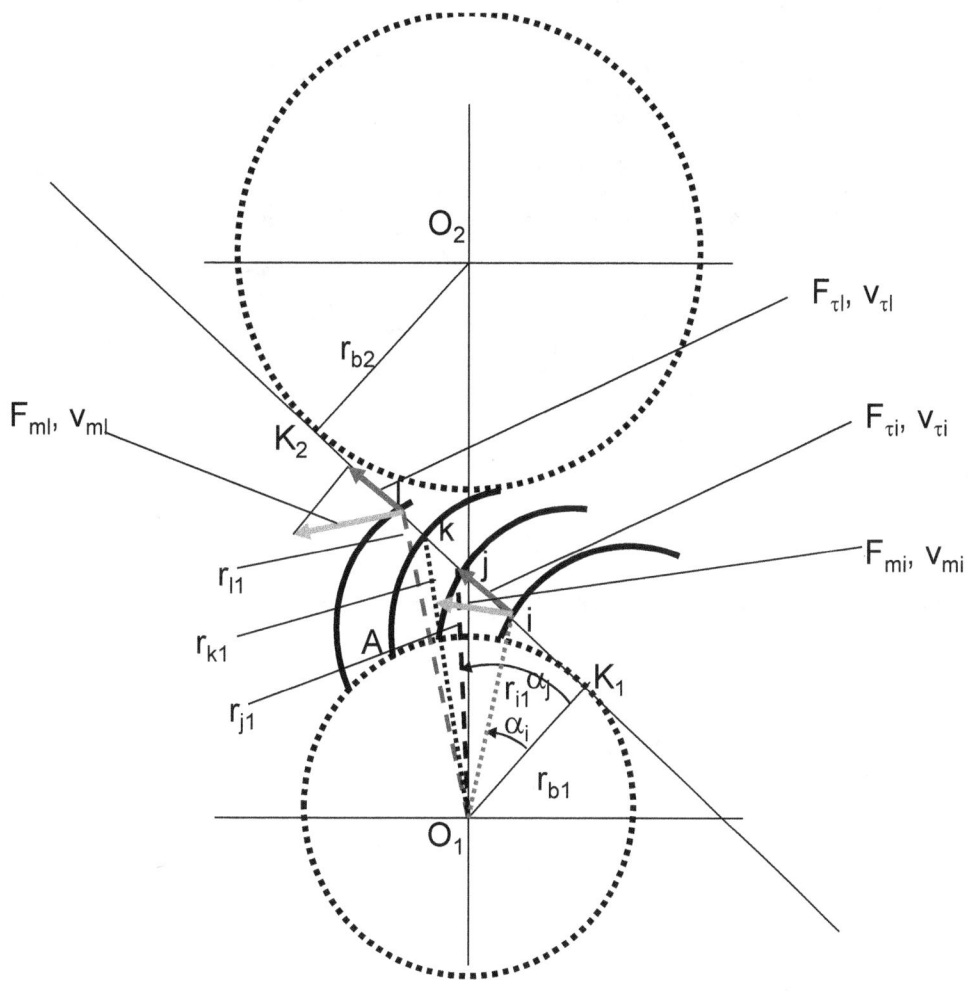

Fig. 1 Four pairs of teeth in contact concomitantly

$$v_{mi} = \frac{r_{b1} \cdot \omega_1}{\cos \alpha_i}; v_{mj} = \frac{r_{b1} \cdot \omega_1}{\cos \alpha_j}; v_{mk} = \frac{r_{b1} \cdot \omega_1}{\cos \alpha_k}; v_{ml} = \frac{r_{b1} \cdot \omega_1}{\cos \alpha_l} \qquad (3)$$

The forces transmitted concomitantly at the four points must be the same (4):

$$F_{\tau i} = F_{\tau j} = F_{\tau k} = F_{\tau l} = F_\tau \qquad (4)$$

The motor forces are (5):

$$F_{mi} = \frac{F_\tau}{\cos \alpha_i}; F_{mj} = \frac{F_\tau}{\cos \alpha_j}; F_{mk} = \frac{F_\tau}{\cos \alpha_k}; F_{ml} = \frac{F_\tau}{\cos \alpha_l} \qquad (5)$$

The momentary efficiency can be written in the form (6).

$$\eta_i = \frac{P_u}{P_c} = \frac{P_\tau}{P_m} = \frac{F_{\tau i} \cdot v_{\tau i} + F_{\tau j} \cdot v_{\tau j} + F_{\tau k} \cdot v_{\tau k} + F_{\tau l} \cdot v_{\tau l}}{F_{mi} \cdot v_{mi} + F_{mj} \cdot v_{mj} + F_{mk} \cdot v_{mk} + F_{ml} \cdot v_{ml}} =$$

$$= \frac{4 \cdot F_\tau \cdot r_{b1} \cdot \omega_1}{\dfrac{F_\tau \cdot r_{b1} \cdot \omega_1}{\cos^2 \alpha_i} + \dfrac{F_\tau \cdot r_{b1} \cdot \omega_1}{\cos^2 \alpha_j} + \dfrac{F_\tau \cdot r_{b1} \cdot \omega_1}{\cos^2 \alpha_k} + \dfrac{F_\tau \cdot r_{b1} \cdot \omega_1}{\cos^2 \alpha_l}} =$$

$$= \frac{4}{\dfrac{1}{\cos^2 \alpha_i} + \dfrac{1}{\cos^2 \alpha_j} + \dfrac{1}{\cos^2 \alpha_k} + \dfrac{1}{\cos^2 \alpha_l}} =$$

$$= \frac{4}{4 + tg^2 \alpha_i + tg^2 \alpha_j + tg^2 \alpha_k + tg^2 \alpha_l} \qquad (6)$$

Relations (7) and (8) are auxiliary relations:

$$\begin{aligned}
& K_1 i = r_{b1} \cdot tg\alpha_i; K_1 j = r_{b1} \cdot tg\alpha_j; \\
& K_1 k = r_{b1} \cdot tg\alpha_k; K_1 l = r_{b1} \cdot tg\alpha_l \\
& K_1 j - K_1 i = r_{b1} \cdot (tg\alpha_j - tg\alpha_i); \\
& K_1 j - K_1 i = r_{b1} \cdot \frac{2 \cdot \pi}{z_1} \Rightarrow tg\alpha_j = tg\alpha_i + \frac{2 \cdot \pi}{z_1} \\
& K_1 k - K_1 i = r_{b1} \cdot (tg\alpha_k - tg\alpha_i); \\
& K_1 k - K_1 i = r_{b1} \cdot 2 \cdot \frac{2 \cdot \pi}{z_1} \Rightarrow tg\alpha_k = tg\alpha_i + 2 \cdot \frac{2 \cdot \pi}{z_1} \\
& K_1 l - K_1 i = r_{b1} \cdot (tg\alpha_l - tg\alpha_i); \\
& K_1 l - K_1 i = r_{b1} \cdot 3 \cdot \frac{2 \cdot \pi}{z_1} \Rightarrow tg\alpha_l = tg\alpha_i + 3 \cdot \frac{2 \cdot \pi}{z_1}
\end{aligned} \qquad (7)$$

$$\begin{cases} tg\alpha_j = tg\alpha_i \pm \dfrac{2 \cdot \pi}{z_1}; \\ tg\alpha_k = tg\alpha_i \pm 2 \cdot \dfrac{2 \cdot \pi}{z_1}; \\ tg\alpha_l = tg\alpha_i \pm 3 \cdot \dfrac{2 \cdot \pi}{z_1} \end{cases} \qquad (8)$$

One keeps relations (8), with the sign plus (+) for the gearing where the drive wheel 1 has external teeth (at the external or internal gearing), and with the sign (-) for the gearing where the drive wheel 1, has internal teeth (the drive wheel is a ring, only for the internal gearing).

The relation of the momentary efficiency (6) uses the auxiliary relations (8) and takes the form (9).

$$\eta_i = \frac{4}{4+tg^2\alpha_i+tg^2\alpha_j+tg^2\alpha_k+tg^2\alpha_l} =$$

$$= \frac{4}{4+tg^2\alpha_i+(tg\alpha_i\pm\frac{2\pi}{z_1})^2+(tg\alpha_i\pm 2\cdot\frac{2\pi}{z_1})^2+(tg\alpha_i\pm 3\cdot\frac{2\pi}{z_1})^2} =$$

$$= \frac{4}{4+4\cdot tg^2\alpha_i+\frac{4\pi^2}{z_1^2}\cdot(0^2+1^2+2^2+3^2)\pm 2\cdot tg\alpha_i\cdot\frac{2\pi}{z_1}\cdot(0+1+2+3)} =$$

$$= \frac{1}{1+tg^2\alpha_i+\frac{4\pi^2}{E\cdot z_1^2}\cdot\sum_{i=1}^{E}(i-1)^2\pm 2\cdot tg\alpha_i\cdot\frac{2\pi}{E\cdot z_1}\cdot\sum_{i=1}^{E}(i-1)} =$$

$$= \frac{1}{1+tg^2\alpha_1+\frac{4\pi^2}{E\cdot z_1^2}\cdot\frac{E\cdot(E-1)\cdot(2\cdot E-1)}{6}\pm\frac{4\pi\cdot tg\alpha_1}{E\cdot z_1}\cdot\frac{E\cdot(E-1)}{2}} =$$

$$= \frac{1}{1+tg^2\alpha_1+\frac{2\pi^2\cdot(E-1)\cdot(2E-1)}{3\cdot z_1^2}\pm\frac{2\pi\cdot tg\alpha_1\cdot(E-1)}{z_1}} = \quad (9)$$

$$= \frac{1}{1+tg^2\alpha_1+\frac{2\pi^2}{3\cdot z_1^2}\cdot(\varepsilon_{12}-1)\cdot(2\cdot\varepsilon_{12}-1)\pm\frac{2\pi\cdot tg\alpha_1}{z_1}\cdot(\varepsilon_{12}-1)}$$

In expression (9) one starts with relation (6) where four pairs are in contact concomitantly, but then one generalizes the expression, replacing the 4 figure (four pairs) with E couples, replacing figure 4 with the E variable, which represents the whole number of the contact ratio +1, and after restricting the sums expressions, we replace the variable E with the contact ratio ε_{12}, as well.

The mechanical efficiency offers more advantages than the momentary efficiency, and will be calculated approximately, by replacing in relation (9) the pressure angle α_1, with the normal pressure angle α_0 the relation taking the form (10); where ε_{12} represents the contact ratio of the gearing, and it will be calculated with expression (11) for the external gearing, and with relation (12) for the internal gearing.

$$\eta_m = \frac{1}{1+tg^2\alpha_0+\frac{2\pi^2}{3\cdot z_1^2}\cdot(\varepsilon_{12}-1)\cdot(2\cdot\varepsilon_{12}-1)\pm\frac{2\pi\cdot tg\alpha_0}{z_1}\cdot(\varepsilon_{12}-1)} \quad (10)$$

$$\varepsilon_{12}^{a.e.} = \frac{\sqrt{z_1^2\cdot\sin^2\alpha_0+4\cdot z_1+4}+\sqrt{z_2^2\cdot\sin^2\alpha_0+4\cdot z_2+4}-(z_1+z_2)\cdot\sin\alpha_0}{2\cdot\pi\cdot\cos\alpha_0} \quad (11)$$

$$\varepsilon_{12}^{a.i.} = \frac{\sqrt{z_e^2\cdot\sin^2\alpha_0+4\cdot z_e+4}-\sqrt{z_i^2\cdot\sin^2\alpha_0-4\cdot z_i+4}+(z_i-z_e)\cdot\sin\alpha_0}{2\cdot\pi\cdot\cos\alpha_0} \quad (12)$$

The calculations made have been centralized in the table 1.

Table 1
The centralized results

z_1	α_0	z_2	ε_{12}^{ae}	η_{12}^{ae}	η_{21}^{ae}	ε_{12}^{ai}	η_{12}^{ai}	η_{21}^{ai}
42	20	126	1.79	0.844	0.871	1.92	0.838	0.895
46	19	138	1.87	0.856	0.882	2.00	0.850	0.905
52	18	156	1.96	0.869	0.893	2.09	0.864	0.915
58	17	174	2.06	0.880	0.904	2.20	0.876	0.925
65	16	195	2.17	0.892	0.914	2.32	0.887	0.933
74	15	222	2.30	0.903	0.923	2.46	0.899	0.942
85	14	255	2.44	0.914	0.933	2.62	0.910	0.949
98	13	294	2.62	0.924	0.941	2.81	0.920	0.956
115	12	345	2.82	0.934	0.949	3.02	0.931	0.963
137	11	411	3.06	0.943	0.957	3.28	0.941	0.969
165	10	495	3.35	0.952	0.964	3.59	0.950	0.974
204	9	510	3.68	0.960	0.970	4.02	0.958	0.980
257	8	514	4.09	0.968	0.975	4.57	0.966	0.985
336	7	672	4.66	0.975	0.980	5.21	0.973	0.989
457	6	914	5.42	0.981	0.985	6.06	0.980	0.992
657	5	1314	6.49	0.986	0.989	7.26	0.986	0.994

3 Determining of gearing efficiency in function of the contact ratio to the bended teeth

Generally we use gearings with teeth inclined (with bended teeth). For gears with bended teeth, the calculations show a decrease in yield when the inclination angle increases. For angles with inclination which not exceed 25 degree the efficiency of gearing is good (see the table 2). When the inclination angle (β) exceeds 25 degrees the gearing will suffer a significant drop in yield (see the tables 3-4).

Table 2. *Bended teeth, $\beta=25$ [deg].*

			Determining the efficiency when $\beta=25$ [deg]					
z_1	α_0 [grad]	z_2	ε_{12}^{ae}	η_{12}^{ae}	η_{21}^{ae}	ε_{12}^{ai}	η_{12}^{ai}	η_{21}^{ai}
42	20	126	1,708	0,829	0,851	1,791	0,826	0,871
46	19	138	1,776	0,843	0,864	1,865	0,839	0,883
52	18	156	1,859	0,856	0,876	1,949	0,853	0,895
58	17	174	1,946	0,869	0,889	2,043	0,866	0,906
65	16	195	2,058	0,882	0,900	2,151	0,879	0,917
74	15	222	2,165	0,894	0,911	2,275	0,892	0,927
85	14	255	2,299	0,906	0,922	2,418	0,904	0,936
98	13	294	2,456	0,917	0,932	2,584	0,915	0,945
115	12	345	2,641	0,928	0,941	2,780	0,926	0,953
137	11	411	2,863	0,938	0,950	3,013	0,937	0,961
165	10	495	3,129	0,948	0,958	3,295	0,947	0,968
204	9	510	3,443	0,957	0,965	3,665	0,956	0,974
257	8	514	3,829	0,965	0,971	4,146	0,964	0,981
336	7	672	4,357	0,973	0,977	4,719	0,972	0,985
457	6	914	5,064	0,980	0,983	5,486	0,979	0,989
657	5	1314	6,056	0,985	0,988	6,563	0,985	0,992

Table 3. *Bended teeth, β=35 [deg].*

Determining the efficiency when $\beta=35\ [deg]$								
z_1	α_0 [grad]	z_2	ε_{12}^{ae}	η_{12}^{ae}	η_{21}^{ae}	ε_{12}^{ai}	η_{12}^{ai}	η_{21}^{ai}
42	20	126	1,620	0,809	0,827	1,677	0,807	0,843
46	19	138	1,681	0,825	0,841	1,741	0,822	0,858
52	18	156	1,755	0,840	0,856	1,815	0,838	0,871
58	17	174	1,832	0,854	0,870	1,898	0,852	0,885
65	16	195	1,948	0,868	0,883	1,993	0,867	0,897
74	15	222	2,030	0,882	0,896	2,103	0,881	0,909
85	14	255	2,150	0,895	0,909	2,230	0,894	0,921
98	13	294	2,293	0,908	0,920	2,379	0,907	0,932
115	12	345	2,461	0,920	0,931	2,554	0,919	0,942
137	11	411	2,663	0,932	0,942	2,764	0,931	0,951
165	10	495	2,906	0,942	0,951	3,017	0,942	0,959
204	9	510	3,196	0,952	0,959	3,345	0,952	0,968
257	8	514	3,556	0,962	0,967	3,766	0,961	0,975
336	7	672	4,041	0,970	0,974	4,281	0,969	0,981
457	6	914	4,692	0,978	0,981	4,971	0,977	0,986
657	5	1314	5,607	0,984	0,986	5,942	0,984	0,990

Table 4. *Bended teeth, β=45 [deg].*

Determining the efficiency when $\beta=45\ [deg]$								
z_1	α_0 [grad]	z_2	ε_{12}^{ae}	η_{12}^{ae}	η_{21}^{ae}	ε_{12}^{ai}	η_{12}^{ai}	η_{21}^{ai}
42	20	126	1,505	0,772	0,784	1,539	0,771	0,796
46	19	138	1,555	0,790	0,802	1,590	0,789	0,814
52	18	156	1,618	0,808	0,820	1,650	0,807	0,831
58	17	174	1,680	0,825	0,837	1,718	0,824	0,848
65	16	195	1,810	0,841	0,853	1,796	0,841	0,864
74	15	222	1,848	0,858	0,869	1,888	0,858	0,879
85	14	255	1,949	0,874	0,884	1,994	0,874	0,894
98	13	294	2,070	0,889	0,899	2,119	0,889	0,908
115	12	345	2,215	0,904	0,913	2,268	0,903	0,921
137	11	411	2,389	0,918	0,926	2,446	0,917	0,933
165	10	495	2,600	0,931	0,938	2,662	0,930	0,944
204	9	510	2,855	0,943	0,948	2,938	0,943	0,955
257	8	514	3,173	0,954	0,958	3,290	0,954	0,965
336	7	672	3,599	0,964	0,967	3,732	0,964	0,973
457	6	914	4,171	0,973	0,976	4,325	0,973	0,980
657	5	1314	4,976	0,981	0,983	5,161	0,981	0,986

New calculation relationships can be put in the forms (13-15).

$$\eta_m = \frac{z_1^2 \cdot \cos^2 \beta}{z_1^2(tg^2\alpha_0 + \cos^2 \beta) + \frac{2}{3}\pi^2 \cos^4 \beta(\varepsilon-1)(2\varepsilon-1) \pm 2\pi tg\alpha_0 z_1 \cos^2 \beta(\varepsilon-1)} \quad (13)$$

$$\varepsilon^{a.e.} = \frac{1+tg^2\beta}{2\cdot\pi} \cdot \left\{ \sqrt{[(z_1+2\cdot\cos\beta)\cdot tg\alpha_0]^2 + 4\cdot\cos^3\beta\cdot(z_1+\cos\beta)} + \sqrt{[(z_2+2\cdot\cos\beta)\cdot tg\alpha_0]^2 + 4\cdot\cos^3\beta\cdot(z_2+\cos\beta)} - (z_1+z_2)\cdot tg\alpha_0 \right\} \quad (14)$$

$$\varepsilon^{a.i.} = \frac{1+tg^2\beta}{2\cdot\pi} \cdot \left\{ \sqrt{[(z_e+2\cdot\cos\beta)\cdot tg\alpha_0]^2 + 4\cdot\cos^3\beta\cdot(z_e+\cos\beta)} - \sqrt{[(z_i-2\cdot\cos\beta)\cdot tg\alpha_0]^2 - 4\cdot\cos^3\beta\cdot(z_i-\cos\beta)} - (z_e-z_i)\cdot tg\alpha_0 \right\} \quad (15)$$

The calculation relationships (13-15) are general. They have the advantage that can be used with great precision in determining the efficiency of any type of gearings.

To use them at the gearing without bended teeth is enough to assign them a beta value = zero. The results obtained in this case will be identical to the ones of the relations 10-12.

4 Conclusions

The best efficiency can be obtained with the internal gearing when the drive wheel 1 is the ring; the minimum efficiency will be obtained when the drive wheel 1 of the internal gearing has external teeth.

For the external gearing, the best efficiency is obtained when the bigger wheel is the drive wheel; *when we decrease the normal angle α_0, the contact ratio increases and the efficiency increases as well.*

The efficiency increases too, when the number of teeth of the drive wheel 1 increases (when z_1 increases).

References

[1] PETRESCU R.V., PETRESCU F.I., POPESCU N., „Determining Gear Efficiency", In Gear Solutions magazine, USA, pp. 19-28, March 2007.

CHAPTER III

PRESENTING A "DYNAMIC ORIGINAL MODEL" USED TO STUDY TOOTHED GEARING WITH PARALLEL AXES

Abstract: Nearly all the models studied the dynamic on gearing with axes parallel, is based on mechanical models of classical (known) who is studying spinning vibration of shafts gears and determine their own beats and strains of shafts spinning; sure that they are very useful, but are not actually join formed of the two teeth in contact (or more pairs of teeth in contact), that is not treated physiology of the mechanism itself with toothed gears for a view that the phenomena are dynamic taking place in top gear flat; model [1] just try this so, but the whole theory is based directly on the impact of teeth (collisions between teeth); this chapter will present an original model that tries to explore the dynamic phenomena taking place in the plane geared couple from the geared transmissions with parallel axes.

Keywords: Gear dynamics, spinning vibration, shaft vibration, impact of teeth, geared couple, angle characteristic

1. Introduction (or the starting idea)

Nearly all the models [1, 2, 3, 7] studied the dynamic on gearing with axes parallel, is based on mechanical models of classical (known) who is studying spinning vibration of shafts gears and determine their own beats and strains of shafts spinning; sure that they are very useful, but are not actually join formed of the two teeth in contact (or more pairs of teeth in contact), that is not treated physiology of the mechanism itself with toothed gears for a view that the phenomena are dynamic taking place in top gear flat; model [1] just try this so, but the whole theory is based directly on the impact of teeth (collisions between teeth); this chapter will present an original model that tries to explore the dynamic phenomena taking place in the plane geared couple from the geared transmissions with parallel axes.

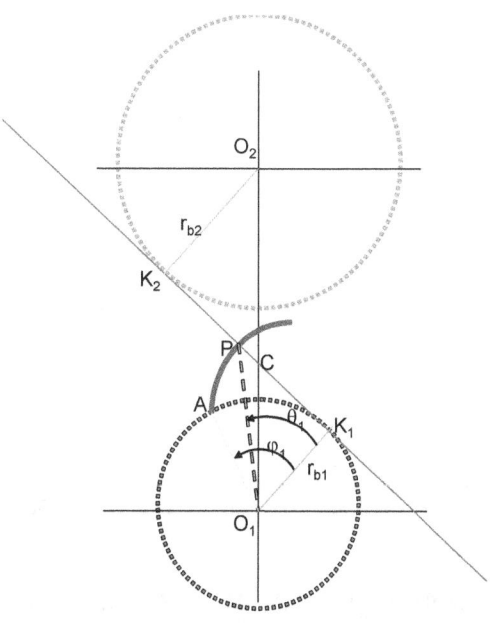

Fig. 1. *Angles characteristic at a position of a tooth from the driving wheel, in gearing*

Figure 1 presents a tooth of the lead wheel 1, in gearing, at a certain position on the gearing segment K_1K_2. It is characterized by angles θ_1 and φ_1, the first showing the position of the vector O_1P (the contact vector) in relation to fixed vector O_1K_1, and the second showing how much is turned the tooth (leading wheel 1) in relation to O_1K_1.

Between the two angles are the relations of liaison 1:

$$\varphi_1 = tg\,\theta_1 \qquad \theta_1 = arctg\,\varphi_1 \qquad (1)$$

Since φ_1 is the sum of angles θ_1 and γ_1, where the angle γ_1 represents the known function $inv\theta_1$:

$$\varphi_1 = \theta_1 + \gamma_1 = \theta_1 + inv\,\theta_1 = \theta_1 + (tg\,\theta_1 - \theta_1) = tg\,\theta_1 \qquad (2)$$

One derives the relations (1) and one obtains the forms (3):

$$\begin{cases} \dot{\varphi}_1 = (1 + tg^2\theta_1)\cdot\dot{\theta}_1 = (1+\varphi_1^2)\cdot\dot{\theta}_1 \\[2mm] \dot{\theta}_1 = \dfrac{\dot{\varphi}_1}{1+\varphi_1^2} = D_1\cdot\omega_1;\; D_1 = \dfrac{1}{1+\varphi_1^2} = \dfrac{1}{1+tg^2\theta_1} \\[4mm] \ddot{\theta}_1 = \dot{D}_1\cdot\omega_1 = D_1'\cdot\omega_1^2;\; D_1' = \dfrac{-2\cdot\varphi_1}{(1+\varphi_1^2)^2} = \dfrac{-2\cdot tg\,\theta_1}{(1+tg^2\theta_1)^2} \end{cases} \qquad (3)$$

2. The dynamic model

The dynamic model considered [3, 4] (fig. 2) is similar to that of cam gears [4, 5, 6], as geared wheels are similar to those with camshaft; practically the toothed wheel is a multiple cam, each tooth is a lobe (cam) showing only the up lifting phase. Forces and J * (M *) is amended, so equation of motion will get another look.

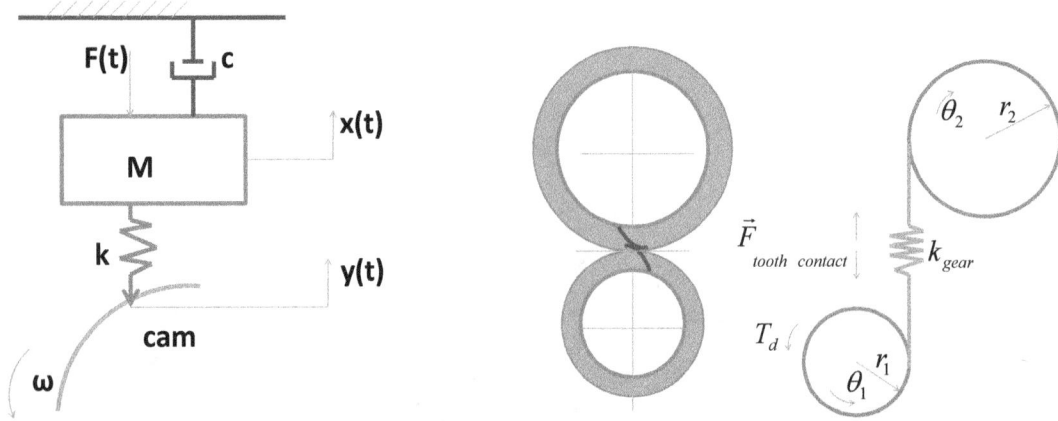

Fig. 2. *The dynamic model. Forces, displacements, and elasticity of the system*

Contact between the two teeth is practically a contact between a rotation cam and a rocker follower. So similar to models with cams [5, 6] it will determine precision cinematic (dynamics cinematic) to join with gears with parallel axes. Vector which should be the leading at the driving wheel 1 (in the dynamics cinematic), is the contact vector O₁P, the angle of his position as θ_1 and his angular velocity, $\dot{\theta}_1$. To the driven wheel 2, one forwards the speed v₂ (see schedule kinematics in Figure 3).

$$v_2 = -v_1 \cdot \cos\theta_1 = -r_{p1} \cdot \dot{\theta}_1 \cdot \cos\theta_1 = -r_{b1} \cdot \dot{\theta}_1 = -r_{b1} \cdot D_1 \cdot \omega_1$$

$$\text{but}: \quad v_2 = r_{b2} \cdot \omega_2 \Rightarrow \omega_2 = -\frac{r_{b1}}{r_{b2}} \cdot D_1 \cdot \omega_1 \tag{4}$$

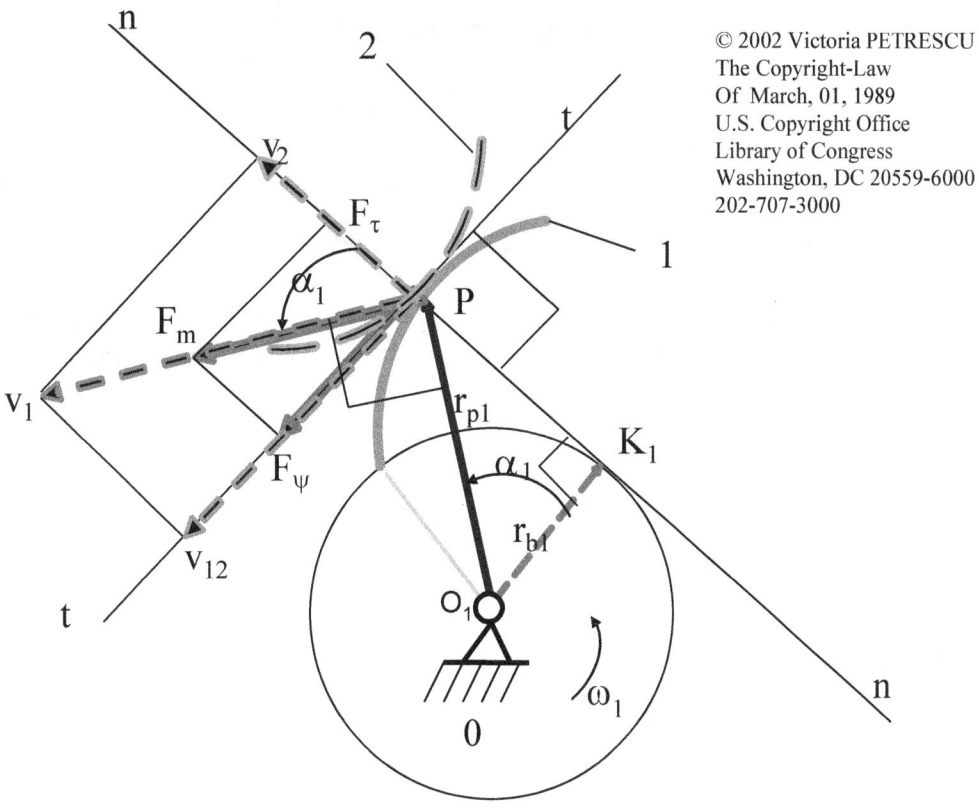

Fig. 3. *Forces and velocities characteristic at a position of a tooth from the driving wheel, in gearing*

By derivation is calculated and angular acceleration (precision acceleration), to the wheel 2 (5), and by integration, movement of the wheel 2 (6):

$$\varepsilon_2 = -\frac{r_{b1}}{r_{b2}} \cdot D_1' \cdot \omega_1^2 \tag{5}$$

$$\varphi_2 = -\frac{r_{b1}}{r_{b2}} \cdot arctg(\varphi_1) = -\frac{r_{b1}}{r_{b2}} \cdot \theta_1 \tag{6}$$

Reduced Force (engines and resistance) at the wheel 1, lead, is equal to the elastic force of couple (while at the wheel led 2 does not intervene and a strong technological force or one additional) and is written in the form (7):

$$F^* = K \cdot (r_{b1} \cdot \varphi_1 - r_{b2} \cdot \varphi_2) =$$
$$= K \cdot (r_{b1} \cdot \varphi_1 - r_{b2} \cdot \frac{r_{b1}}{r_{b2}} \cdot \theta_1) = \quad (7)$$
$$= K \cdot r_{b1} \cdot (tg\theta_1 - \theta_1)$$

Minus sign was already taken, so φ_2 is replaced just in module in the expression of 7, and K means constant elastic of teeth in contact, and is measured in [N/m]. Dynamic equation of motion is written:

$$M^* \cdot \ddot{x} + \frac{1}{2} \cdot \frac{dM^*}{dt} \cdot \dot{x} = F^* \quad (8)$$

Reduced mass, M^*, is determined by the relationship (9):

$$M^* = (J_1 + \frac{1}{i^2} \cdot J_2) \cdot \frac{1}{r_{p1}^2} = (J_1 + \frac{1}{i^2} \cdot J_2) \cdot \frac{\cos^2 \theta_1}{r_{b1}^2} =$$
$$= \frac{J_1 + \frac{1}{i^2} \cdot J_2}{r_{b1}^2} \cdot \cos^2 \theta_1 = C_M \cdot \cos^2 \theta_1 ; C_M = \frac{J_1 + \frac{1}{i^2} \cdot J_2}{r_{b1}^2} \quad (9)$$

Where, J_1 and J_2 represent the moments of inertia (mass, mechanical), reduced to wheel 1, while i is the module of transmission ratio from the wheel 1 to the wheel 2 (see relation 10):

$$i = \frac{r_{b2}}{r_{b1}} = -\frac{\omega_1}{\omega_2} \quad (10)$$

The moving x of the wheel 2 on the gearing segment, writes:

$$x = r_{b2} \cdot \varphi_2 = r_{b2} \cdot -\frac{r_{b1}}{r_{b2}} \cdot arctg\varphi_1 = -r_{b1} \cdot arctg\varphi_1 = -r_{b1} \cdot \theta_1 \quad (11)$$

The speed and corresponding accelerations can be written:

$$\dot{x} = -r_{b1} \cdot \dot{\theta}_1 = -r_{b1} \cdot \frac{1}{1+tg^2\theta_1} \cdot \omega_1 \quad (12)$$

$$\ddot{x} = -r_{b1} \cdot \ddot{\theta}_1 = -r_{b1} \cdot \frac{-2 \cdot tg\theta_1}{(1+tg^2\theta_1)^2} \cdot \omega_1^2 = 2 \cdot r_{b1} \cdot \frac{tg\theta_1}{(1+tg^2\theta_1)^2} \cdot \omega_1^2 \quad (13)$$

One derives the reduced mass and result the expression (14):

$$\frac{dM^*}{dt} = -2 \cdot C_M \cdot \frac{\cos\theta_1 \cdot \sin\theta_1}{1+tg^2\theta_1} \cdot \omega_1 \quad (14)$$

The equation of motion (8) takes now the form (15), which can arrange and form (16):

$$3 \cdot C_M \cdot r_{b1} \cdot \frac{tg\theta_1}{(1+tg^2\theta_1)^3} \cdot \omega_1^2 = K \cdot r_{b1} \cdot (tg\theta_1 - \theta_1) \quad (15)$$

$$\theta_1^d \equiv \theta_1 = tg\theta_1 - \frac{3 \cdot C_M \cdot tg\theta_1 \cdot \omega_1^2}{K \cdot (1+tg^2\theta_1)^3} =$$
$$= tg\theta_1 \cdot [1 - \frac{3 \cdot (J_1 + \frac{r_{b1}^2}{r_{b2}^2} \cdot J_2) \cdot \omega_1^2}{r_{b1}^2 \cdot K \cdot (1+tg^2\theta_1)^3}] \quad (16)$$

The expression (16) is the solution equation of motion of superior couple; to an angle of rotation of the wheel 1, φ_1, known, which is corresponding a pressure angle θ_1 known, the expression (16) generates a dynamic angle of pressure, θ_1^d.

In terms of the constant elasticity of the teeth in contact, K, is large enough, if the radius of the base circle of the wheel 1 don't decrease too much (z_1 to be greater than 15-20), for normal speeds and even higher (but not too large), the ratio of expression parenthesis 16 remains under the value 1, and even much lower than 1, and the expression 16 can be engineering estimated to the natural short form (17):

$$\theta_1^d = tg\theta_1 = \varphi_1^c \equiv \varphi_1 \tag{17}$$

3. The (dynamic) angular velocity at the lead wheel 1

We can determine now the instantaneous (momentary) angular velocity of the lead wheel 1 (relationship 19); it used the intermediate relation (18) as well:

$$\frac{\Delta\omega_1}{\omega_m} = \frac{\Delta\varphi_1}{\varphi_1} \Rightarrow \Delta\omega_1 = \frac{\Delta\varphi_1}{\varphi_1}\cdot\omega_m = \frac{\varphi_1^d - \varphi_1}{\varphi_1}\cdot\omega_m = \\ = \frac{tg(\theta_1^d) - \varphi_1}{\varphi_1}\cdot\omega_m = \frac{tg(\varphi_1) - \varphi_1}{\varphi_1}\cdot\omega_m = \frac{inv\varphi_1}{\varphi_1}\cdot\omega_m \tag{18}$$

$$\omega_1 = \omega_m + \Delta\omega_1 = (1 + \frac{inv\varphi_1}{\varphi_1})\cdot\omega_m = \\ = \frac{tg(\varphi_1)}{\varphi_1}\cdot\omega_m = \frac{tg(tg\theta_1)}{tg\theta_1}\cdot\omega_m = R_{d1}\cdot\omega_m \tag{19}$$

It defines the dynamic coefficient, R_{d1}, as the ratio between the tangent of the angle φ_1 and φ_1 angle, or the ratio, $\frac{tg(tg\theta_1)}{tg\theta_1}$, relationship 1.20:

$$R_{d1} = \frac{tg(tg\theta_1)}{tg\theta_1} \tag{20}$$

Dynamic synthesis of gearing with axes parallel can be made taking into account the relation (1.20). The necessity of obtaining a dynamic factor as low (close to the value 1), requires limiting the maximum pressure angle, θ_{1M} and the normal angle α_0, and increasing the minimum number of teeth of the leading wheel, 1, z_{1min}.

4. The dynamic of wheel 2 (conducted)

One can determine now the instantaneous (momentary) dynamic angular velocity at the led wheel 2 (relationship 28), and all angular parameters (displacement, velocity, acceleration), in three situations: classical cinematic, precision cinematic, and dynamic; where: c=cinematic, cp=precision cinematic, d=dynamic (see the relation: 21-31).

$$\varphi_2^c = -\frac{r_{b1}}{r_{b2}} \cdot \varphi_1 \tag{21}$$

$$\omega_2^c = -\frac{r_{b1}}{r_{b2}} \cdot \omega_1 \tag{22}$$

$$\varepsilon_2^c = -\frac{r_{b1}}{r_{b2}} \cdot \varepsilon_1 = 0 \tag{23}$$

$$\varphi_2^{cp} = -\frac{r_{b1}}{r_{b2}} \cdot arctg\,\varphi_1 = -\frac{r_{b1}}{r_{b2}} \cdot \theta_1 \tag{24}$$

$$\omega_2^{cp} = -\frac{r_{b1}}{r_{b2}} \cdot \frac{1}{1+\varphi_1^2} \cdot \omega_1 = -\frac{r_{b1}}{r_{b2}} \cdot \frac{1}{1+tg^2\theta_1} \cdot \omega_1 \tag{25}$$

$$\varepsilon_2^{cp} = -\frac{r_{b1}}{r_{b2}} \cdot \frac{-2\cdot\varphi_1}{(1+\varphi_1^2)^2} \cdot \omega_1^2 = -\frac{r_{b1}}{r_{b2}} \cdot \frac{-2\cdot tg\,\theta_1}{(1+tg^2\theta_1)^2} \cdot \omega_1^2 \tag{26}$$

Dynamics of wheel 2 (conducted) are calculated with relations (27-31):

$$\varphi_2^d = -\frac{r_{b1}}{r_{b2}} \cdot \int \frac{tg\,\varphi_1}{\varphi_1 + \varphi_1^3} d\varphi_1 \tag{27}$$

$$\omega_2^d = -\frac{r_{b1}}{r_{b2}} \cdot \frac{1}{1+\varphi_1^2} \cdot \frac{tg\,\varphi_1}{\varphi_1} \cdot \omega_1 = -\frac{r_{b1}}{r_{b2}} \cdot \frac{1}{1+tg^2\theta_1} \cdot \frac{tg(tg\,\theta_1)}{tg\,\theta_1} \cdot \omega_1 \tag{28}$$

$$\varepsilon_2^d = -\frac{r_{b1}}{r_{b2}} \cdot \frac{(1+tg^2\varphi_1)\cdot(\varphi_1+\varphi_1^3) - tg\,\varphi_1\cdot(1+3\cdot\varphi_1^2)}{(\varphi_1+\varphi_1^3)^2} \cdot \omega_1^2 \tag{29}$$

With:

$$\varphi_{1m} = tg\,\theta_{1m} = \frac{(z_1+z_2)\cdot\sin\alpha_0 - \sqrt{z_2^2\cdot\sin^2\alpha_0 + 4\cdot z_2 + 4}}{z_1\cdot\cos\alpha_0} \tag{30}$$

$$\varphi_{1M} = tg\,\theta_{1M} = \frac{\sqrt{z_1^2\cdot\sin^2\alpha_0 + 4\cdot z_1 + 4}}{z_1\cdot\cos\alpha_0} \tag{31}$$

Can be defined to wheel 2 a dynamic coefficient Rd2, (see the relations 28 and 32):

$$R_{d2} = \frac{1}{1+\varphi_1^2} \cdot \frac{tg\,\varphi_1}{\varphi_1} = \frac{1}{1+tg^2\theta_1} \cdot \frac{tg(tg\,\theta_1)}{tg\,\theta_1} \tag{32}$$

5. Conclusions

Representation of angular velocity, ω_2, depending on the angle φ_1, for r_{b1} and r_{b2} known (z_1, z_2, m, and α_0 imposed), and for a certain amount of angular velocity input constant (imposed by the speed of the shaft which is mounted wheel leading 1), can be seen in the figures 4: a, b.

Observe the appearance of vibration at the dynamic angular velocity, ω_2.

Start with equal rays and 20 degrees all for, α_0 (fig. 4a), and stay on the chart last rays equal and α_0 reduced to 5 degrees (fig. 4b).

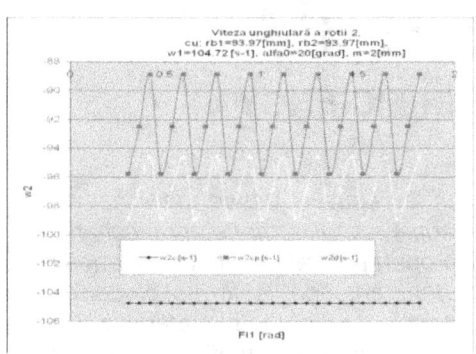 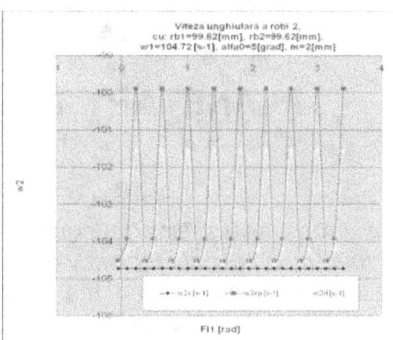

Fig. 4. *Dynamics of wheel 2; ω_2 cinematic, ω_2 in precision cinematic, ω_2 dynamic*

In fig. 5 it can see the experimental vibration [1, 2].

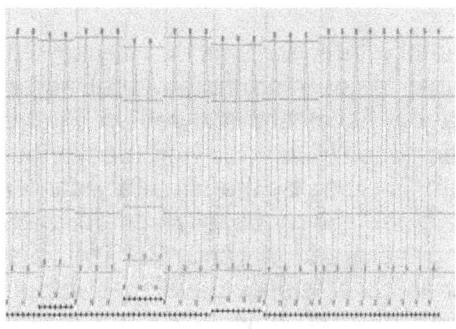

Fig. 5. *Dynamics of wheel 2; ω_2 dynamic (the vibrations obtained experimental)*

The presented (original) method is more simply, directly, naturaly and rapidly than the classics.

References

[1]-Bajer, A., *„Parallel finite element simulator of planetary gear trains"*, In Ph. D. Dissertation, The University of Texas, 2001;

[2]-Li, J., *„Gear Fatigue Crack Prognosis Using Embedded Model, Gear Dynamic Model and Fracture Mechanics"*, Department of Mechanical, Aerospace and Nuclear Engineering, Rensselaer Polytechnic Institute;

[3]-Peeters, J., Vandepitte, P., *„Flexible multibody model of a three-stage planetary gear-box in a wind turbine"*, In Proceedings of ISMA, 2004, p. 3923-3942;

[4]-PETRESCU, F.I., PETRESCU, R.V. *Contributions at the dynamics of cams.* In the Ninth IFToMM International Symposium on Theory of Machines and Mechanisms, SYROM 2005, Bucharest, Romania, 2005, Vol. I, p. 123-128;

[5]-PETRESCU, R.V., COMĂNESCU A., PETRESCU F.I., *Dynamics of Cam Gears Illustrated at the Classic Distribution Mechanism.* In NEW TRENDS IN MECHANISMS, Ed. Academica - Greifswald, 2008, I.S.B.N. 978-3-9402-37-10-1;

[6]-PETRESCU, R.V., COMĂNESCU A., PETRESCU F.I., ANTONESCU O., *The Dynamics of Cam Gears at the Module B (with Translated Follower with Roll).* In NEW TRENDS IN MECHANISMS, Ed. Academica - Greifswald, 2008, I.S.B.N. 978-3-9402-37-10-1.

[7]-Takeuchi, T., Togai, K., *„Gear Whine Analysis with Virtual Power Train"*, In Mitsubishi Motors Technical Review, 2004, No. 16, p. 23-28.

[8]-PETRESCU F.I., PETRESCU R.V., *Presenting A Dynamic Original Model Used to Study Toothed Gearing with Parallel Axes.* In the 13-th edition of SCIENTIFIC CONFERENCE with international participation, Constantin Brâncusi University, Târgu-Jiu, November 2008.

CHAPTER IV
PLANETARY TRAINS EFFICIENCY

Abstract: *Synthesis of classical planetary mechanisms is usually based on kinematic relations, considering in especially the transmission ratio input-output achieved. The planetary mechanisms are less synthesized based by their mechanical efficiency developed in operation, although this criterion is part of the real dynamics of mechanisms, being also the most important criterion in terms of performance of a mechanism. Even when it used the efficiency criterion, the determination of the planetary yield, is made only with approximate relationships. The most widely recognized method is one method of Russian school of mechanisms. This chapter determines the exact method to calculate the mechanical efficiency of a planetary mechanism. In this mode it is resolving one important problem of the dynamics of planetary mechanisms.*
Keywords: *planetary efficiency, planetary kinematic, planetary synthesis*

1. Introduction

Synthesis of classical planetary mechanisms is usually based on kinematic relations, considering in especially the transmission ratio input-output achieved. The most common model used is the differential planetary mechanism showed in Figure 1. Usually the formula 1 is determined by writing the relationship Willis (1'). For the various cinematic planetary systems presented in Figure 3, where entry is made by the planetary carrier (H), and output is achieved by the final element (f), the initial element being usually immobilized, will be used for the kinematic calculations the relationships generalized 1 and 2. The planetary mechanisms are less synthesized based by their mechanical efficiency developed in operation, although this criterion is part of the real dynamics of mechanisms, being also the most important criterion in terms of performance of a mechanism. Even when it used the efficiency criterion, the determination of the planetary yield, is made only with approximate relationships. The most widely recognized method is one method of Russian school of mechanisms. This chapter determines the real efficiency of the planetary trains (determines the exact method).

2. Kinematic synthesis

Synthesis of classical planetary mechanisms is usually based on kinematic relations, considering in especially the transmission ratio input-output achieved. The most common model used is the differential planetary mechanism showed in the Figure 1.

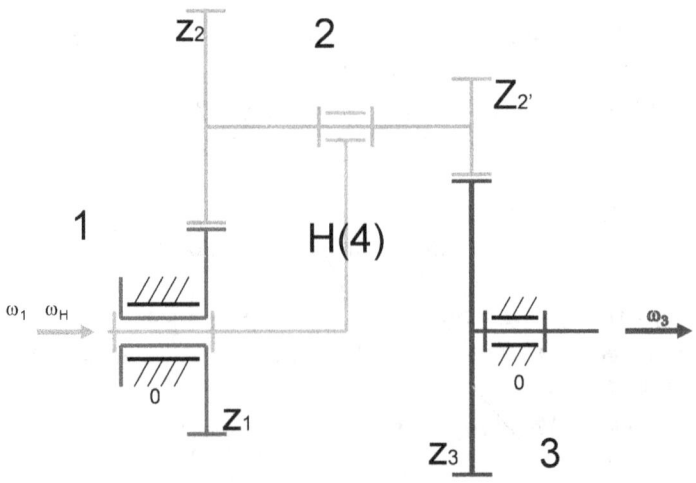

Fig. 1. *Kinematic schema of a differential planetary mechanism (M=2)*

For this mechanism to have one single degree of mobility, remaining in use with a drive desmodromic unique and single output, it is necessary to reduce the mobility of the mechanism from two to one, which can be obtained by connecting in series or parallel of two or more planetary gear, by binding to gears with fixed axes, or the hardening of one of its mobile elements; element 1 in this case (case in which the wheel 1 is identified with the fixed element 0; fig. 2).

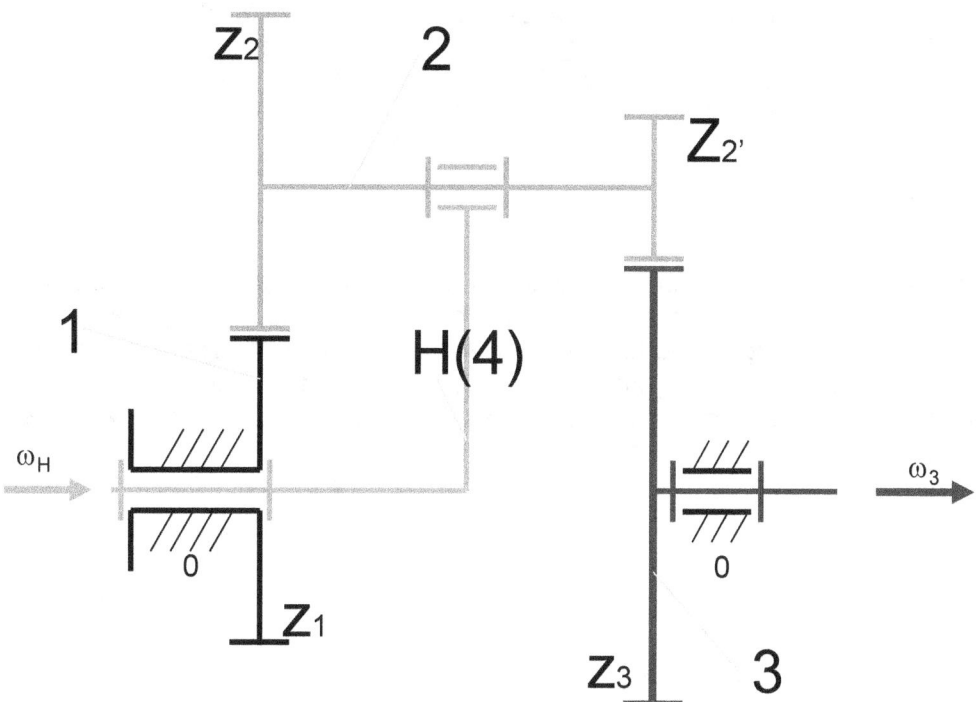

Fig. 2. *Kinematic schema of a simple planetary mechanism (M=1)*

The entrance to the simple planetary from Figure 2 is made by the planetary carrier (H), and the output is done by the mobile cinematic element (3), the wheel (3). Kinematic ratio between input-output (H-3) is written directly with the relationship 1.

$$i_{H3}^{1} = \frac{1}{i_{3H}^{1}} = \frac{1}{1 - i_{31}^{H}} = \frac{1}{1 - \frac{1}{i_{13}^{H}}} \qquad (1)$$

Where i_{13}^{H} is the ratio of transmission input output corresponding to the mechanism with fixed axis (when the planetary carrier H is fixed), and is determined in function of the cinematic schema of planetary gear used; for the model in Figure 2 it is determined by the relation 2, as a depending on the numbers of teeth of the wheels 1, 2, 2 ', 3.

$$i_{13}^{H} = \frac{z_2}{z_1} \cdot \frac{z_3}{z_{2'}} \qquad (2)$$

Usually the formula 1 is determined by writing the relationship Willis (1').

$$\begin{cases} i_{13}^H = \dfrac{\omega_1 - \omega_H}{\omega_3 - \omega_H} \equiv \dfrac{z_2}{z_1} \cdot \dfrac{z_3}{z_{2'}} \\[2ex] \dfrac{z_2}{z_1} \cdot \dfrac{z_3}{z_{2'}} = \dfrac{\dfrac{\omega_1}{\omega_H} - \dfrac{\omega_H}{\omega_H}}{\dfrac{\omega_3}{\omega_H} - \dfrac{\omega_H}{\omega_H}} \\[3ex] i_{13}^H = \dfrac{z_2 \cdot z_3}{z_1 \cdot z_{2'}} = \dfrac{0-1}{\dfrac{\omega_3}{\omega_H}-1} = \dfrac{1}{1 - i_{3H}} = \dfrac{1}{1 - \dfrac{1}{i_{H3}^1}} \Rightarrow \\[3ex] \Rightarrow i_{H3}^1 = \dfrac{1}{1 - \dfrac{1}{i_{13}^H}} \end{cases} \quad (1') \qquad i_{Hf}^i = \dfrac{1}{i_{fH}^i} = \dfrac{1}{1 - i_{fi}^H} = \dfrac{1}{1 - \dfrac{1}{i_{if}^H}} \quad (3)$$

For the various cinematic planetary systems presented in Figure 3, where entry is made by the planetary carrier (H), and output is achieved by the final element (f), the initial element being usually immobilized, will be used for the kinematic calculations the relationships generalized 1 and 2; the relationship 1 takes the general form 3, and the relation 2 is written in one of the forms 4 particularized for each schema separately, used; where i become 1, and f takes the value 3 or 4 as appropriate.

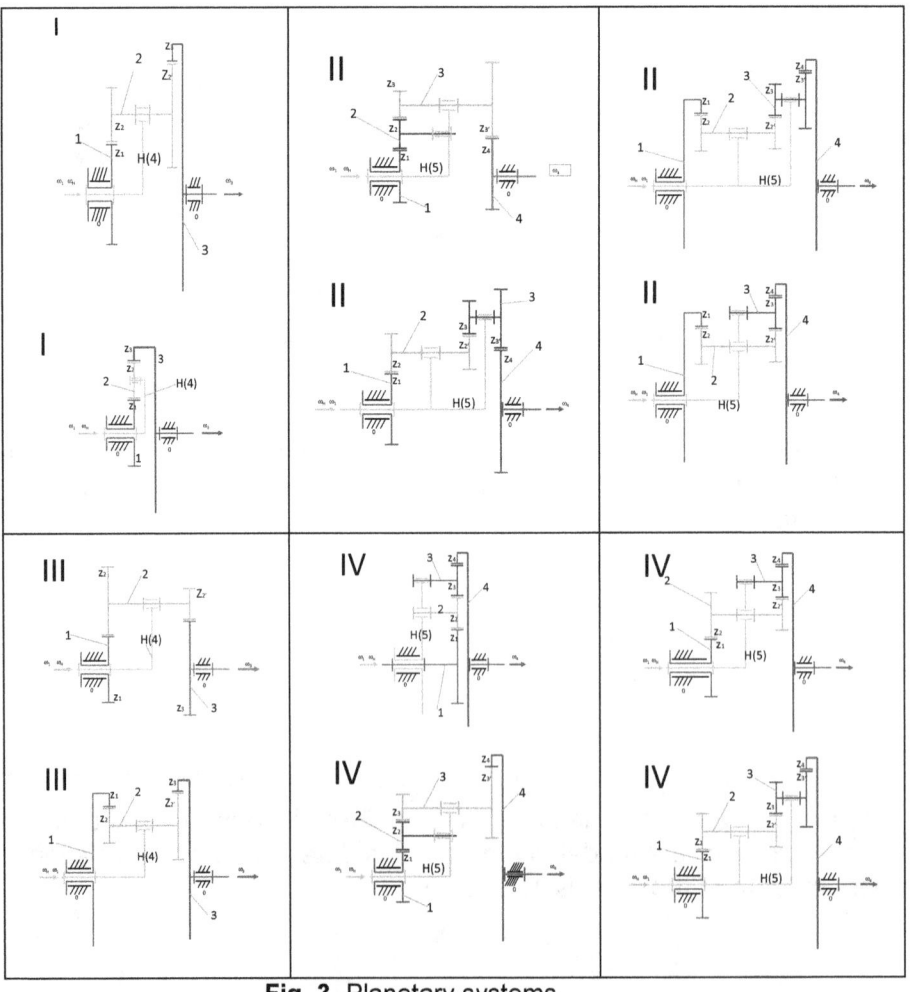

Fig. 3. Planetary systems

$$\begin{cases} i_{13}^H = -\dfrac{z_2}{z_1} \cdot \dfrac{z_3}{z_{2'}} & \text{for } I \text{ of up} \\[6pt] i_{13}^H = -\dfrac{z_3}{z_1} & \text{for } I \text{ of down} \\[6pt] i_{13}^H = \dfrac{z_2}{z_1} \cdot \dfrac{z_3}{z_{2'}} & \text{for } III \text{ of up} \\[6pt] i_{13}^H = \dfrac{z_2}{z_1} \cdot \dfrac{z_3}{z_{2'}} & \text{for } III \text{ of down} \\[6pt] i_{14}^H = -\dfrac{z_3}{z_1} \cdot \dfrac{z_4}{z_{3'}} & \text{for } II \text{ left up} \\[6pt] i_{14}^H = -\dfrac{z_2}{z_1} \cdot \dfrac{z_3}{z_{2'}} \cdot \dfrac{z_4}{z_{3'}} & \text{for } II \text{ right up} \\[6pt] i_{14}^H = -\dfrac{z_2}{z_1} \cdot \dfrac{z_3}{z_{2'}} \cdot \dfrac{z_4}{z_{3'}} & \text{for } II \text{ left down} \\[6pt] i_{14}^H = -\dfrac{z_2}{z_1} \cdot \dfrac{z_4}{z_{2'}} & \text{for } II \text{ right down} \\[6pt] i_{14}^H = \dfrac{z_4}{z_1} & \text{for } IV \text{ left up} \\[6pt] i_{14}^H = \dfrac{z_2}{z_1} \cdot \dfrac{z_4}{z_{2'}} & \text{for } IV \text{ right up} \\[6pt] i_{14}^H = \dfrac{z_3}{z_1} \cdot \dfrac{z_4}{z_{3'}} & \text{for } IV \text{ left down} \\[6pt] i_{14}^H = \dfrac{z_2}{z_1} \cdot \dfrac{z_3}{z_{2'}} \cdot \dfrac{z_4}{z_{3'}} & \text{for } IV \text{ right down} \end{cases} \quad (4)$$

The planetary mechanisms are less synthesized based by their mechanical efficiency developed in operation, although this criterion is part of the real dynamics of mechanisms, being also the most important criterion in terms of performance of a mechanism.

Even when it used the efficiency criterion, the determination of the planetary yield, is made only with approximate relationships.

The most widely recognized method is one method of Russian school of mechanisms.

This chapter determines the real efficiency of the planetary trains (the exact method).

The automatic transmissions have been added slowly from airplanes to automobiles, and were then generalized to various vehicles. By using the formulas indicated in this paper for calculating the dynamic of planetary mechanisms, planetary trains, and planetary systems, used in aircraft and vehicles, their automatic transmissions can be achieved better than those known today.

In this mode it is resolving one important problem of the dynamics of planetary mechanisms. For a complete dynamic, may need to be determined and the dynamics deformation mechanisms (see the Figure 4). But this is not part of the subject matter of this book.

Fig. 4. *The deformation of a planetary mechanism*

3. Dynamic synthesis, based on performance achieved

Dynamic synthesis of planetary trains based on performance achieved, can be made with the relationships presented below.

For a normally planetary system (Figure 2) the mechanical efficiency is determined by starting from the relationship 5, which gives the lost power (P_l) in function of the input power (P_H) and the output power (P_3 or P_4, and generic P_f).

$$\begin{aligned}
P_l &= P_H - P_3 = M_H \cdot \omega_H - M_3 \cdot \omega_3 = \\
&= (M_3 + M_1) \cdot \omega_H - M_3 \cdot \omega_3 = \\
&= M_3 \cdot \omega_H - M_3 \cdot \omega_3 + M_1 \cdot \omega_H = \\
&= M_3 \cdot (\omega_H - \omega_3) + M_1 \cdot \omega_H
\end{aligned} \tag{5}$$

It is known the Willis relationship (6), from that it can explicit the moment (M_1). With M_1 put in the relationship (5) it obtains the expression (7).

$$\begin{cases}
\eta_{13}^H = \dfrac{P_3^H}{P_1^H} = \dfrac{M_3 \cdot \omega_3^H}{M_1 \cdot \omega_1^H} = \dfrac{M_3 \cdot (\omega_3 - \omega_H)}{M_1 \cdot (\omega_1 - \omega_H)} = \\
= \dfrac{M_3}{M_1} \cdot \dfrac{\omega_3 - \omega_H}{-\omega_H} = \dfrac{M_3}{M_1} \cdot \left(1 - \dfrac{\omega_3}{\omega_H}\right) = \\
= \dfrac{M_3}{M_1} \cdot (1 - i_{3H}) = \dfrac{M_3}{M_1} \cdot (1 - i_{3H}^1) \Rightarrow \\
\Rightarrow M_1 = \dfrac{M_3}{\eta_{13}^H} \cdot (1 - i_{3H}^1)
\end{cases} \tag{6}$$

$$\begin{cases} P_l = M_3 \cdot (\omega_H - \omega_3) + M_1 \cdot \omega_H = \\ = M_3 \cdot (\omega_H - \omega_3) + \dfrac{M_3 \cdot \omega_H}{\eta_{13}^H} \cdot (1 - i_{3H}) = \\ = M_3 \cdot \omega_3 \cdot \left(\dfrac{\omega_H}{\omega_3} - 1\right) + M_3 \cdot \omega_3 \cdot \left(\dfrac{\omega_H}{\omega_3} - 1\right) \cdot \dfrac{1}{\eta_{13}^H} = \\ = M_3 \cdot \omega_3 \cdot \left(\dfrac{\omega_H}{\omega_3} - 1\right) \cdot \left(1 + \dfrac{1}{\eta_{13}^H}\right) = \\ = M_3 \cdot \omega_3 \cdot (i_{H3} - 1) \cdot \dfrac{1 + \eta_{13}^H}{\eta_{13}^H} = P_3 \cdot (i_{H3} - 1) \cdot \dfrac{1 + \eta_{13}^H}{\eta_{13}^H} \\ \Rightarrow P_p = |P_l| = P_3 \cdot \dfrac{1 + \eta_{13}^H}{\eta_{13}^H} \cdot |i_{H3} - 1| \end{cases} \quad (7)$$

The exact efficiency can be obtained by putting the expression of the "real lost power Pp" (the absolute lost power, determined by the relationship 7) in the formula of the mechanical efficiency of one planetary system (8).

$$\begin{cases} \eta_{H3}^1 = \dfrac{P_3}{P_H} = \dfrac{P_3}{P_3 + P_p} = \dfrac{P_3}{P_3 + P_3 \cdot \dfrac{1 + \eta_{13}^H}{\eta_{13}^H} \cdot |i_{H3} - 1|} = \\ = \dfrac{1}{1 + \dfrac{1 + \eta_{13}^H}{\eta_{13}^H} \cdot |i_{H3} - 1|} = \dfrac{1}{1 + \dfrac{1 + \eta_{13}^H}{\eta_{13}^H} \cdot |i_{H3}^1 - 1|} \end{cases} \quad (8)$$

For the mechanisms with four toothed wheels the efficiency takes the form (9).

$$\begin{cases} \eta_{H4}^1 = \dfrac{P_4}{P_H} = \dfrac{P_4}{P_4 + P_p} = \dfrac{P_4}{P_4 + P_4 \cdot \dfrac{1 + \eta_{14}^H}{\eta_{14}^H} \cdot |i_{H4} - 1|} = \\ = \dfrac{1}{1 + \dfrac{1 + \eta_{14}^H}{\eta_{14}^H} \cdot |i_{H4} - 1|} = \dfrac{1}{1 + \dfrac{1 + \eta_{14}^H}{\eta_{14}^H} \cdot |i_{H4}^1 - 1|} \end{cases} \quad (9)$$

4. Conclusions

The planetary system efficiency given by the exact formula is less than the calculated efficiency by known classical formulas.

If these new relationships are correct, the planetary systems work generally with lower efficiency.

References

[1] Petrescu, F.I., Petrescu, R.V., *Planetary Trains*, Book, LULU Publisher, USA, April 2011, ISBN 978-1-4476-0696-3, pages 204.

CHAPTER V

CAM GEARS EFFICIENCY

Abstract: *The chapter presents an original method to determine the efficiency of a mechanism with cam and follower. The originality of this method consists in eliminating the friction modulus. In this chapter it analyses four types of cam mechanisms: 1.The mechanism with rotary cam and plate translated follower; 2.The mechanism with rotary cam and translated follower with roll; 3.The mechanism with rotary cam and rocking-follower with roll; 4.The mechanism with rotary cam and plate rocking-follower. For every kind of cam and follower mechanism one uses a different method to determine the most efficient design. We take into account the cam's mechanism (distribution mechanism), which is the second mechanism in internal-combustion engines. The optimizing of this mechanism (the distribution mechanism), can improve the functionality of the engine and may increase the comfort of the vehicle too.*

Keywords: *cam, efficiency, translated follower, rocking-follower, follower with roll*

1 Introduction

In this chapter the authors present an original method to calculate the efficiency of the cam's mechanisms. Four kinds of cam and follower mechanisms are analyzed: 1. A mechanism with rotary cam and plate translated follower; 2. A mechanism with rotary cam and translated follower with roll; 3. A mechanism with rotary cam and rocking-follower with roll; 4. A mechanism with rotary cam and plate rocking follower. For every kind of cams and followers mechanism, a different method for the cam's design with a better efficiency has been utilized.

2 Determining of momentary mechanical efficiency of the rotary cam and plate translated follower

The consumed motor force, F_c, perpendicular at A to the vector r_A, is divided into two components [1, 2]: a) F_m, which represents the useful force, or the motor force reduced to the follower; b) F_ψ, which is the sliding force between the two profiles of cam and follower (Fig. 1). See the written relations (2.1-2.10):

$$F_m = F_c \cdot \sin \tau \qquad (2.1)$$

$$v_2 = v_1 \cdot \sin \tau \qquad (2.2)$$

$$P_u = F_m \cdot v_2 = F_c \cdot v_1 \cdot \sin^2 \tau \qquad (2.3)$$

$$P_c = F_c \cdot v_1 \qquad (2.4)$$

$$\eta_i = \frac{P_u}{P_c} = \frac{F_c \cdot v_1 \cdot \sin^2 \tau}{F_c \cdot v_1} = \sin^2 \tau = \cos^2 \delta \qquad (2.5)$$

$$\sin^2 \tau = \frac{s'^2}{r_A^2} = \frac{s'^2}{(r_0+s)^2 + s'^2} \qquad (2.6)$$

$$F_\psi = F_c \cdot \cos \tau \qquad (2.7)$$

$$v_{12} = v_1 \cdot \cos \tau \qquad (2.8)$$

$$P_\psi = F_\psi \cdot v_{12} = F_c \cdot v_1 \cdot \cos^2 \tau \qquad (2.9)$$

$$\psi_i = \frac{P_\psi}{P_c} = \frac{F_c \cdot v_1 \cdot \cos^2 \tau}{F_c \cdot v_1} = \cos^2 \tau = \sin^2 \delta \qquad (2.10)$$

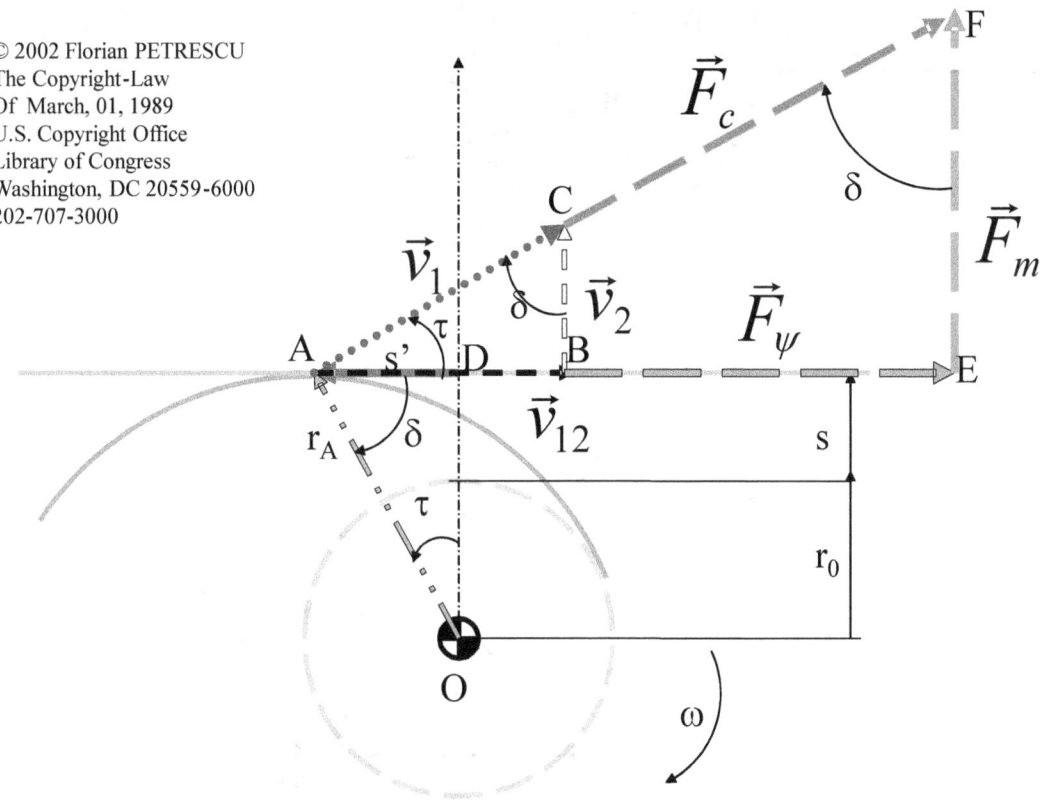

© 2002 Florian PETRESCU
The Copyright-Law
Of March, 01, 1989
U.S. Copyright Office
Library of Congress
Washington, DC 20559-6000
202-707-3000

Fig. 1 *Forces and speeds to the cam with plate translated follower*

3 Determining of momentary dynamic efficiency of the rotary cam and translated follower with roll

The pressure angle δ (Fig. 2), is determined by relations (3.5-3.6) [1, 2]. We can write the next forces, speeds and powers (3.13-3.18). F_m, v_m, are perpendicular to the vector r_A at A. F_m is divided into F_a (the sliding force) and F_n (the normal force). F_n is divided too, into F_i (the bending force) and F_u (the useful force). The momentary dynamic efficiency can be obtained from relation (3.18):

The written relations are the following.

$$r_B^2 = e^2 + (s_0 + s)^2 \qquad (3.1)$$

$$r_B = \sqrt{r_B^2} \qquad (3.2)$$

$$\cos\alpha_B \equiv \sin\tau = \frac{e}{r_B} \qquad (3.3)$$

$$\sin\alpha_B \equiv \cos\tau = \frac{s_0 + s}{r_B} \quad (3.4)$$

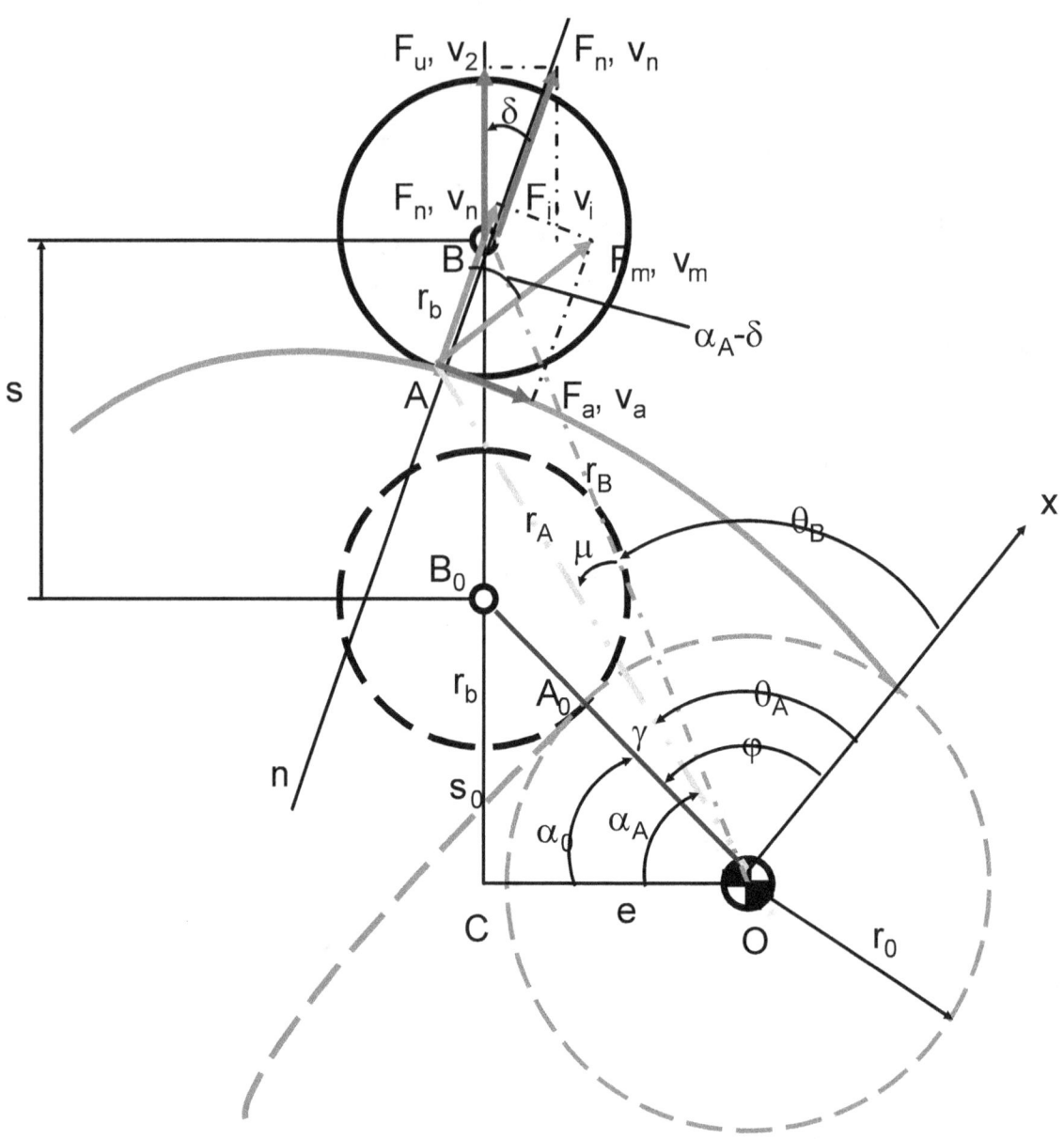

Fig. 2 *Forces and speeds to the cam with translated follower with roll*

$$\cos\delta = \frac{s_0 + s}{\sqrt{(s_0 + s)^2 + (s'-e)^2}} \quad (3.5)$$

$$\sin\delta = \frac{s'-e}{\sqrt{(s_0 + s)^2 + (s'-e)^2}} \quad (3.6)$$

$$\cos(\delta + \tau) = \cos\delta \cdot \cos\tau - \sin\delta \cdot \sin\tau \quad (3.7)$$

$$r_A^2 = r_B^2 + r_b^2 - 2 \cdot r_b \cdot r_B \cdot \cos(\delta + \tau) \quad (3.8)$$

$$\cos\alpha_A = \frac{e\cdot\sqrt{(s_0+s)^2+(s'-e)^2}+r_b\cdot(s'-e)}{r_A\cdot\sqrt{(s_0+s)^2+(s'-e)^2}} \quad (3.9)$$

$$\sin\alpha_A = \frac{(s_0+s)\cdot[\sqrt{(s_0+s)^2+(s'-e)^2}-r_b]}{r_A\cdot\sqrt{(s_0+s)^2+(s'-e)^2}} \quad (3.10)$$

$$\cos(\alpha_A-\delta) = \frac{(s_0+s)\cdot s'}{r_A\cdot\sqrt{(s_0+s)^2+(s'-e)^2}} = \frac{s'}{r_A}\cdot\cos\delta \quad (3.11)$$

$$\cos(\alpha_A-\delta)\cdot\cos\delta = \frac{s'}{r_A}\cdot\cos^2\delta \quad (3.12)$$

$$\begin{cases} v_a = v_m\cdot\sin(\alpha_A-\delta) \\ F_a = F_m\cdot\sin(\alpha_A-\delta) \end{cases} \quad (3.13)$$

$$\begin{cases} v_n = v_m\cdot\cos(\alpha_A-\delta) \\ F_n = F_m\cdot\cos(\alpha_A-\delta) \end{cases} \quad (3.14)$$

$$\begin{cases} v_i = v_n\cdot\sin\delta \\ F_i = F_n\cdot\sin\delta \end{cases} \quad (3.15)$$

$$\begin{cases} v_2 = v_n\cdot\cos\delta = v_m\cdot\cos(\alpha_A-\delta)\cdot\cos\delta \\ F_u = F_n\cdot\cos\delta = F_m\cdot\cos(\alpha_A-\delta)\cdot\cos\delta \end{cases} \quad (3.16)$$

$$\begin{cases} P_u = F_u\cdot v_2 = F_m\cdot v_m\cdot\cos^2(\alpha_A-\delta)\cdot\cos^2\delta \\ P_c = F_m\cdot v_m \end{cases} \quad (3.17)$$

$$\eta_i = \frac{P_u}{P_c} = \frac{F_m\cdot v_m\cdot\cos^2(\alpha_A-\delta)\cdot\cos^2\delta}{F_m\cdot v_m} =$$
$$= [\cos(\alpha_A-\delta)\cdot\cos\delta]^2 = [\frac{s'}{r_A}\cdot\cos^2\delta]^2 = \frac{s'^2}{r_A^2}\cdot\cos^4\delta \quad (3.18)$$

4 Determining of momentary dynamic efficiency of the rotary cam and rocking follower with roll

F_m, v_m, are perpendicular to the vector r_A at A. F_m is divided into F_a (the sliding force) and F_n (the normal force). F_n is divided too into F_c (the compressed force) and F_u (the useful force). The written relations are the following [1, 2] (4.1-4.31).

$$\cos\psi_0 = \frac{b^2+d^2-(r_0+r_b)^2}{2\cdot b\cdot d} \quad (4.1)$$

$$\psi_2 = \psi+\psi_0 \quad (4.2)$$

$$RAD = \sqrt{d^2+b^2(1-\psi')^2-2bd(1-\psi')\cos\psi_2} \quad (4.3)$$

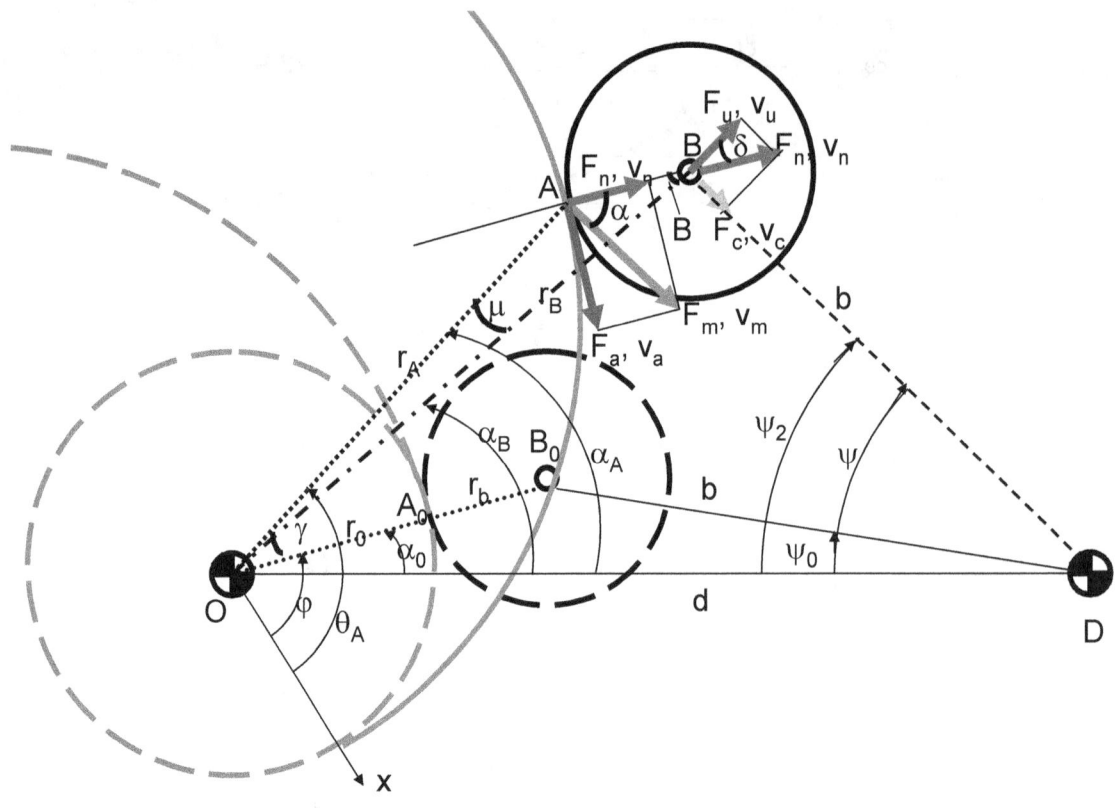

Fig. 3 *Forces and speeds for the rotary cam and rocking follower with roll*

$$\sin\delta = \frac{d\cdot\cos\psi_2 + b\cdot\psi' - b}{RAD} \tag{4.4}$$

$$\cos\delta = \frac{d\cdot\sin\psi_2}{RAD} \tag{4.5}$$

$$r_B^2 = b^2 + d^2 - 2\cdot b\cdot d\cdot\cos\psi_2 \tag{4.6}$$

$$\cos\alpha_B = \frac{d^2 + r_B^2 - b^2}{2\cdot d\cdot r_B} \tag{4.7}$$

$$\sin\alpha_B = \frac{b\cdot\sin\psi_2}{r_B} \tag{4.8}$$

$$\sin(\delta+\psi_2) = \sin\delta\cos\psi_2 + \sin\psi_2\cos\delta \tag{4.9}$$

$$\cos(\delta+\psi_2) = \cos\delta\cos\psi_2 - \sin\psi_2\sin\delta \tag{4.10}$$

$$B = \delta + \psi_2 + \alpha_B - \frac{\pi}{2} \tag{4.11}$$

$$\cos B = \sin(\delta+\psi_2+\alpha_B) \tag{4.12}$$

$$\sin B = -\cos(\delta+\psi_2+\alpha_B) \tag{4.13}$$

$$\cos B = \sin(\delta+\psi_2)\cdot\cos\alpha_B + \sin\alpha_B\cdot\cos(\delta+\psi_2) \tag{4.14}$$

$$\sin B = \sin(\delta + \psi_2) \cdot \sin \alpha_B - \cos \alpha_B \cdot \cos(\delta + \psi_2) \qquad (4.15)$$

$$r_A^2 = r_B^2 + r_b^2 - 2 \cdot r_b \cdot r_B \cdot \cos B \qquad (4.16) \qquad \cos \mu = \frac{r_A^2 + r_B^2 - r_b^2}{2 \cdot r_A \cdot r_B} \qquad (4.17)$$

$$\sin \mu = \frac{r_b}{r_A} \cdot \sin B \qquad (4.18) \qquad \alpha_A = \alpha_B + \mu \qquad (4.19)$$

$$\cos \alpha_A = \cos \alpha_B \cos \mu - \sin \alpha_B \sin \mu \qquad (4.20)$$

$$\sin \alpha_A = \sin \alpha_B \cos \mu + \cos \alpha_B \sin \mu \qquad (4.21)$$

$$\alpha = \pi - \alpha_A - \psi_2 - \delta \qquad (4.22) \qquad \begin{aligned}\cos \alpha &= -\cos(\psi_2 + \delta + \alpha_A) = \\ &= \sin(\psi_2 + \delta) \cdot \sin \alpha_A - \cos(\psi_2 + \delta) \cdot \cos \alpha_A\end{aligned} \qquad (4.23)$$

$$\cos \alpha = \frac{\psi' \cdot b}{r_A} \cdot \cos \delta \qquad (4.24) \qquad \cos \alpha \cdot \cos \delta = \frac{\psi' \cdot b}{r_A} \cdot \cos^2 \delta \qquad (4.25)$$

$$\begin{cases} F_a = F_m \cdot \sin \alpha \\ v_a = v_m \cdot \sin \alpha \end{cases} \qquad (4.26) \qquad \begin{cases} F_n = F_m \cdot \cos \alpha \\ v_n = v_m \cdot \cos \alpha \end{cases} \qquad (4.27)$$

$$\begin{cases} F_c = F_n \cdot \sin \delta \\ v_c = v_n \cdot \sin \delta \end{cases} \qquad (4.28) \qquad \begin{cases} F_u = F_n \cdot \cos \delta = F_m \cdot \cos \alpha \cdot \cos \delta \\ v_2 = v_n \cdot \cos \delta = v_m \cdot \cos \alpha \cdot \cos \delta \end{cases} \qquad (4.29)$$

$$\begin{cases} P_u = F_u \cdot v_2 = F_m \cdot v_m \cdot \cos^2 \alpha \cdot \cos^2 \delta \\ P_c = F_m \cdot v_m \end{cases} \qquad (4.30) \qquad \begin{aligned}\eta_i &= \frac{P_u}{P_c} = \cos^2 \alpha \cdot \cos^2 \delta = (\cos \alpha \cdot \cos \delta)^2 = \\ &= (\frac{\psi' \cdot b}{r_A} \cdot \cos^2 \delta)^2 = \frac{\psi'^2 \cdot b^2}{r_A^2} \cdot \cos^4 \delta\end{aligned} \qquad (4.31)$$

5 Determining of momentary mechanical efficiency of the rotary cam and general plate rocking follower

The written relations are following, (5.1-5.6) (see Fig. 4) [1, 2]:

$$AH = [\sqrt{d^2 - (r_0 - b)^2} \cdot \cos \psi - (r_0 - b) \cdot \sin \psi] \cdot \frac{\psi'}{1 - \psi'} \qquad (5.1)$$

$$OH = b + (r_0 - b) \cdot \cos \psi + \sqrt{d^2 - (r_0 - b)^2} \cdot \sin \psi \qquad (5.2)$$

$$r^2 = AH^2 + OH^2 \qquad (5.3)$$

$$\sin \tau = \frac{AH}{r}; \quad \sin^2 \tau = \frac{AH^2}{r^2} = \frac{AH^2}{AH^2 + OH^2} \qquad (5.4)$$

$$\begin{cases} F_n = F_m \cdot \cos \alpha = F_m \cdot \sin \tau; \\ v_n = v_m \cdot \cos \alpha = v_m \cdot \sin \tau \end{cases} \qquad (5.5)$$

$$\eta_i = \frac{P_n}{P_c} = \frac{F_n \cdot v_n}{F_m \cdot v_m} = \frac{F_m \cdot v_m \cdot \sin^2 \tau}{F_m \cdot v_m} = \sin^2 \tau = \frac{AH^2}{AH^2 + OH^2} \qquad (5.6)$$

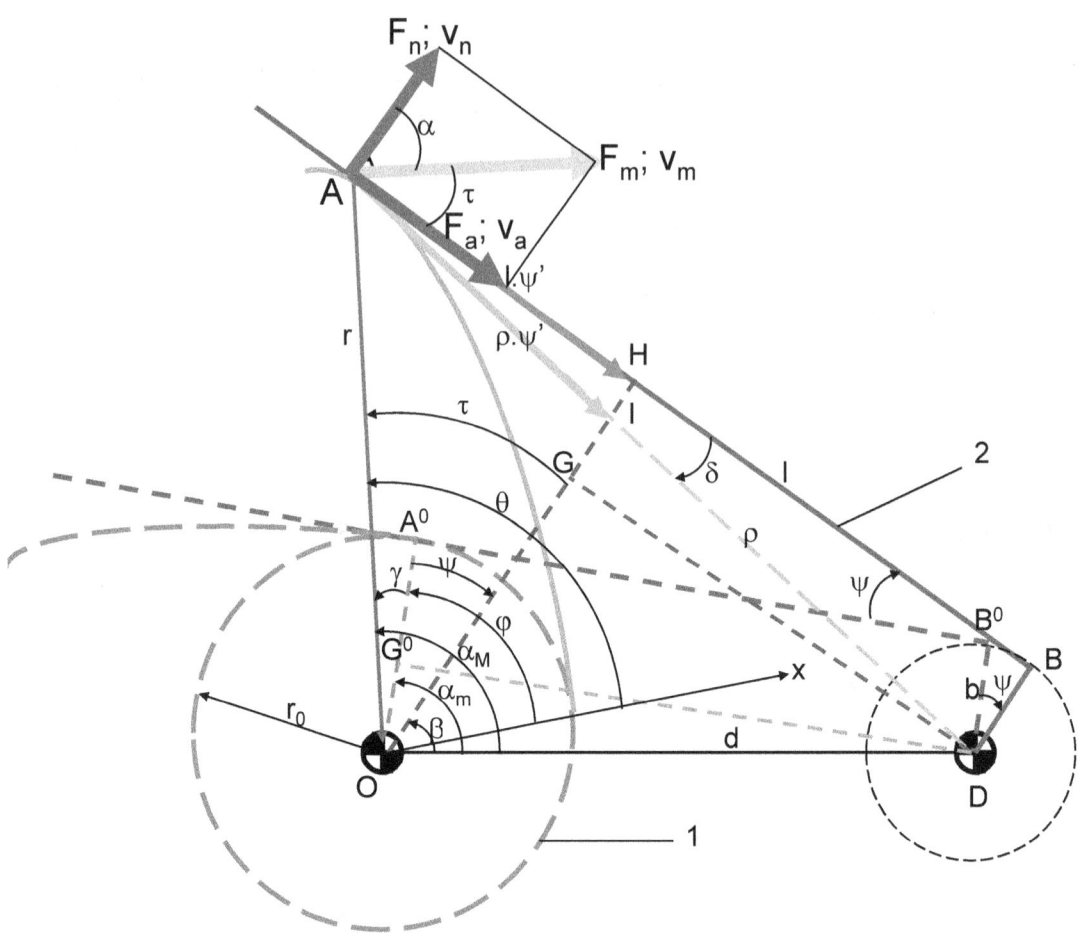

Fig. 4 *Forces and speeds for the rotary cam and general plate rocking follower*

6 Conclusions

The follower with roll makes the input-force be divided into several components. This is the reason why, the dynamics and the precise-kinematics (the dynamic-kinematics) of mechanism with rotary cam and follower with roll, are more different and difficult. The presented dynamic efficiency of followers with roll is not the same like the classical mechanical efficiency. For plate followers the dynamic and the mechanical efficiency are the same. This is the great advantage of plate followers.

References

[1] PETRESCU F.I., PETRESCU R.V., Determining the dynamic efficiency of cams. SYROM 2005, Bucharest, Romania, Vol. I, pp. 129-134, 2005.
[2] PETRESCU F.I., PETRESCU R.V., POPESCU N., The efficiency of cams. In the Second International Conference "Mechanics and Machine Elements", Technical University of Sofia, November 4-6, Sofia, Bulgaria, Vol. II, pp. 237-243, 2005.

CHAPTER VI

CONTRIBUTIONS AT THE DYNAMIC OF CAMS

ABSTRACT: *The chapter presents an original method in determining a general, dynamic and differential equation for the motion of machines and mechanisms, particularized for the mechanisms with rotation cams and followers. This equation can be directly integrated by an original method presented in this chapter. After integration the resulted mother equation may be solved immediately. It presents an original dynamic model with one degree of freedom, with variable internal amortization. It determines the resistant force reduced at the valve (4), the motor force reduced at the valve (5), and the coefficient of variable internal amortization (6). The reduced mass can be calculated with the form (8). The differential motion equation takes the exact form (31), and the approximate form (32). The equation (31) is preparing for its integration with the form (35, 36, 37). The (37) form can be directly integrated and it obtains the parental equation (38). The equation (38) can be arranged in forms (39, 40, 41). The mother equation (41) can be solved directly (42-45), or more elegant with finished differences (48 and 49-50).*

Keywords: *Motor-force, resistant-force, variable internal amortization, differential equation, valve rocker, valve push rod, valve lifter, valve spring.*

1. INTRODUCTION

The chapter presents shortly an original method in determining a general dynamic differential equation, particularized for the mechanisms with rotation cams and followers [1, 2, 3].

This equation can be integrated directly by an original method presented in this chapter.

2. PRESENTING A DYNAMIC MODEL, WITH ONE GRADE OF FREEDOM, WITH VARIABLE INTERNAL AMORTIZATION

2.1. Determining the amortization coefficient of the mechanism

Starting with the kinematical schema of the classical valve gear mechanism (see the figure 1), one creates the translating dynamic model, with a single degree of freedom (with a single mass), with variable internal amortization (see the picture 2), having the motion equation (1).

The formula (1) is just a Newton equation, where the sum of forces on a single element is 0, [1, 2, 3]:

$$M \cdot \ddot{x} = K \cdot (y-x) - k \cdot x - c \cdot \dot{x} - F_0 \qquad (1)$$

Where:

M –the mass of the mechanism, reduced at the valve;

K –the elastically constant of the system;

k –the elastically constant of the valve spring;

c –the coefficient of the system's amortization;

F_0 –the elastically force which compressing the valve spring;

x –the effective displacement of the valve;

y≡s –the theoretical displacement of the tappet reduced at the valve, imposed by the cam's profile.

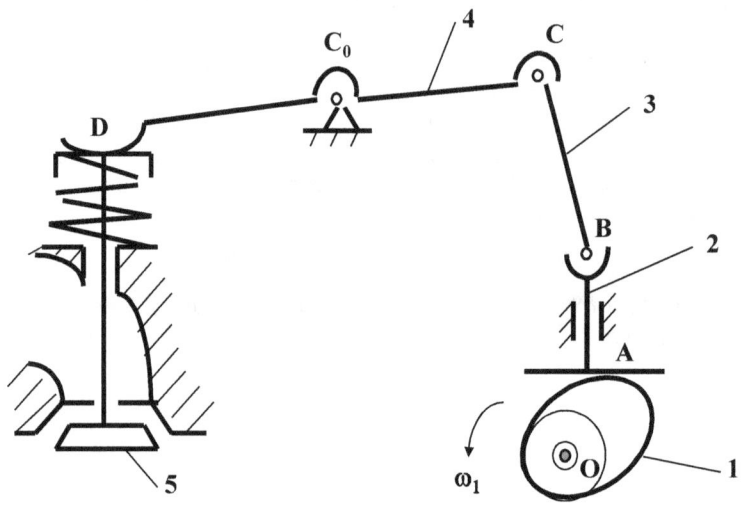

FIG. 1. *The kinematical schema of the classical valve gear mechanism*

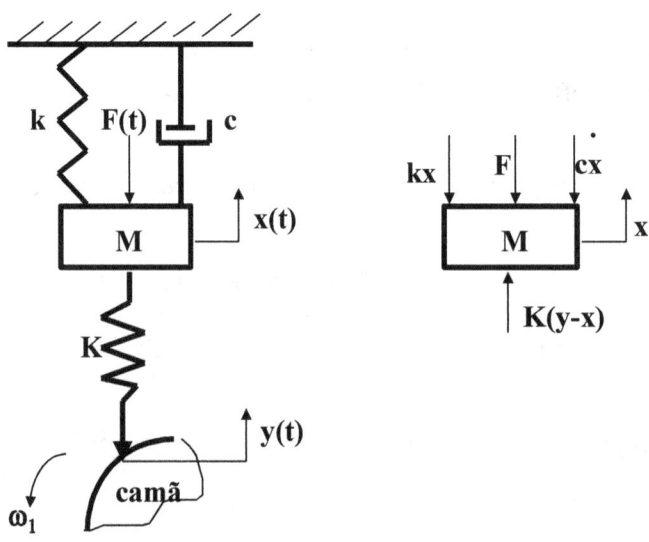

Fig. 2. *Dynamic model with a single liberty, with variable internal amortization*

The Newton equation (1) can be written in form (2):

$$M \cdot \ddot{x} + c \cdot \dot{x} = K \cdot (y - x) - (F_0 + k \cdot x) \tag{2}$$

The differential equation, Lagrange, can be written in form (3).

$$M \cdot \ddot{x} + \frac{1}{2} \cdot \frac{dM}{dt} \cdot \dot{x} = F_m - F_r \tag{3}$$

Comparing the two equations, (2 and 3), we identifie the coefficients and obtain the resistant force (4), the motor force (5) and the coefficient of internal amortization (6), [1, 3]. It can see that the internal amortization coefficient, c, is a variable:

$$F_r = F_0 + k \cdot x = k \cdot x_0 + k \cdot x = k \cdot (x_0 + x) \tag{4}$$

$$F_m = K \cdot (y - x) = K \cdot (s - x) \tag{5}$$

$$c = \frac{1}{2} \cdot \frac{dM}{dt} \tag{6}$$

It places the variable coefficient, c, (see the relation 6), in the Newton equation (form 1 or 2) and obtains the equation (7), [1, 3]:

$$M \cdot \ddot{x} + \frac{1}{2} \cdot \frac{dM}{dt} \cdot \dot{x} + (K + k) \cdot x = K \cdot y - F_0 \tag{7}$$

The reduced mass can be written in form (8), (the reduced mass of the system, reduced at the valve), [1]:

$$M = m_5 + (m_2 + m_3) \cdot \left(\frac{\dot{y}_2}{\dot{x}}\right)^2 + J_1 \cdot \left(\frac{\omega_1}{\dot{x}}\right)^2 + J_4 \cdot \left(\frac{\omega_4}{\dot{x}}\right)^2 \tag{8}$$

With the following notations:

m_2 = the mass of the tappet (of the valve lifter);

m_3 = the mass of the valve push rod;

m_5 = the valve mass;

J_1 = the inertia mechanical moment of the cam;

J_4 = the inertia mechanical moment of the valve rocker;

$\dot{y}_2$ = the tappet velocity, or the second movement-low, imposed by the cam's profile;

$\dot{x}$ = the real (dynamic) valve velocity.

If one notes with $i = i_{25}$, the ratio of transmission tappet-valve, given from the valve rocker, the theoretically velocity of the valve, $\dot{y}$, (the tappet velocity reduced at the valve), takes the form (9), where the ratio of transmission, i, is given from the formula (10).

$$\dot{y} \equiv \dot{y}_5 = \frac{\dot{y}_2}{i} \tag{9}$$

$$i = \frac{CC_0}{C_0 D} \tag{10}$$

It can write the following relations (11-16), where y' is the reduced velocity forced at the tappet by the cam's profile. With the relations (10, 13, 14, 16) the reduced mass (8), can be written in the forms (17–19):

$$\dot{x} = \omega_1 \cdot x' \tag{11}$$

$$\ddot{x} = \omega_1^2 \cdot x'' \tag{12}$$

$$\dot{y}_2 = \omega_1 \cdot y_2' = \omega_1 \cdot i \cdot y' \tag{13}$$

$$\frac{\omega_1}{\dot{x}} = \frac{\omega_1}{\omega_1 \cdot x'} = \frac{1}{x'} \tag{14}$$

$$\omega_4 = \frac{\dot{y}_2}{CC_0} = \frac{\omega_1 \cdot y_2'}{CC_0} = \frac{\omega_1 \cdot y' i}{CC_0} = \frac{\omega_1 \cdot y'}{CC_0} \cdot \frac{CC_0}{C_0 D} = \frac{\omega_1 \cdot y'}{C_0 D} \tag{15}$$

$$\frac{\omega_4}{\dot{x}} = \frac{\omega_1 \cdot y'}{C_0 D \cdot \omega_1 \cdot x'} = \frac{1}{C_0 D} \cdot \frac{y'}{x'} \tag{16}$$

$$M = m_5 + (m_2 + m_3) \cdot (\frac{i \cdot y'}{x'})^2 + J_1 \cdot (\frac{1}{x'})^2 + J_4 \cdot (\frac{1}{C_0 D} \cdot \frac{y'}{x'})^2 \tag{17}$$

$$M = m_5 + [i^2 \cdot (m_2 + m_3) + \frac{J_4}{(C_0 D)^2}] \cdot (\frac{y'}{x'})^2 + J_1 \cdot (\frac{1}{x'})^2 \tag{18}$$

$$M = m_5 + m^* \cdot (\frac{y'}{x'})^2 + J_1 \cdot (\frac{1}{x'})^2 \tag{19}$$

It derivates dM/dφ and obtains the relations (20–22):

$$\frac{d[(\frac{y'}{x'})^2]}{d\varphi} = \frac{2 \cdot y'}{x'} \cdot \frac{(y'' \cdot x' - x'' \cdot y')}{x'^2} =$$
$$= \frac{2 \cdot y'}{x'^2} \cdot (y'' - x'' \cdot \frac{y'}{x'}) = 2 \cdot (\frac{y'}{x'})^2 \cdot (\frac{y''}{y'} - \frac{x''}{x'}) \tag{20}$$

$$\frac{d[(\frac{1}{x'})^2]}{d\varphi} = \frac{2}{x'} \cdot \frac{-x''}{x'^2} = -2 \cdot \frac{x''}{x'^3} \tag{21}$$

$$\frac{dM}{d\varphi} = 2 \cdot m^* \cdot (\frac{y'}{x'})^2 \cdot (\frac{y''}{y'} - \frac{x''}{x'}) - 2 \cdot J_1 \cdot \frac{x''}{x'^3} \tag{22}$$

The relation (6) can be written in form (23) and with relation (22), it's taking the forms (24–25):

$$c = \frac{\omega}{2} \cdot \frac{dM}{d\varphi} \tag{23}$$

$$c = \omega \cdot \{[i^2 \cdot (m_2 + m_3) + \frac{J_4}{(C_0 D)^2}] \cdot (\frac{y'}{x'})^2 \cdot (\frac{y''}{y'} - \frac{x''}{x'}) - J_1 \cdot \frac{x''}{x'^3}\} \tag{24}$$

$$c = \omega \cdot [m^* \cdot (\frac{y'}{x'})^2 \cdot (\frac{y''}{y'} - \frac{x''}{x'}) - J_1 \cdot \frac{x''}{x'^3}] \tag{25}$$

With the notation (26):

$$m^* = i^2 \cdot (m_2 + m_3) + \frac{J_4}{(C_0 D)^2} \tag{26}$$

2.2. Determining the movement equations

With the relations (19, 12, 25, 11) the equation (2) takes the forms (27, 28, 29, 30 and 31):

$$M \cdot \omega^2 \cdot x'' + c \cdot \omega \cdot x' + (K + k) \cdot x = K \cdot y - F_0 \tag{27}$$

$$\omega^2 \cdot x'' \cdot m_5 + \omega^2 \cdot m^* \cdot (\frac{y'}{x'})^2 \cdot x'' + J_1 \cdot (\frac{1}{x'})^2 \cdot x'' \cdot \omega^2 + \omega^2 \cdot x' \cdot m^* \cdot$$
$$(\frac{y'}{x'})^2 \cdot (\frac{y''}{y'} - \frac{x''}{x'}) - x' \cdot \omega^2 \cdot J_1 \cdot \frac{x''}{x'^3} + (K+k) \cdot x = K \cdot y - F_0 \qquad (28)$$

$$\omega^2 \cdot m_5 \cdot x'' + \omega^2 \cdot m^* \cdot x'' \cdot (\frac{y'}{x'})^2 - \omega^2 \cdot m^* \cdot (\frac{y'}{x'})^2 \cdot x'' +$$
$$+ \omega^2 \cdot m^* \cdot y'' \cdot \frac{y'}{x'} + (K+k) \cdot x = K \cdot y - F_0 \qquad (29)$$

$$\omega^2 \cdot m_5 \cdot x'' + (K+k) \cdot x + \omega^2 \cdot m^* \cdot y'' \cdot \frac{y'}{x'} = K \cdot y - F_0 \qquad (30)$$

$$\omega^2 \cdot (m_5 \cdot x'' + m^* \cdot y'' \cdot \frac{y'}{x'}) + (K+k) \cdot x = K \cdot y - F_0 \qquad (31)$$

The exact equation (31) can be approximated at the form (32) with x'≅y':

$$\omega^2 \cdot (m_5 \cdot x'' + m^* \cdot y'') + (K+k) \cdot x = K \cdot y - F_0 \qquad (32)$$

With the following notations: y=s, y'=s', y''=s'', y'''=s''', the equation (32) takes the approximate form (33) and the complete equation (31) takes the exact form (34).

$$\omega^2 \cdot (m_5 \cdot x'' + m^* \cdot s'') + (K+k) \cdot x = K \cdot s - F_0 \qquad (33)$$

$$\omega^2 \cdot (m_5 \cdot x'' + m^* \cdot s'' \cdot \frac{s'}{x'}) + (K+k) \cdot x = K \cdot s - F_0 \qquad (34)$$

3. SOLVING THE DIFFERENTIAL EQUATION BY DIRECT INTEGRATION AND OBTAINING THE MOTHER EQUATION

It integrates the equation (31) directly. It prepares the equation (31) for the integration. First, we write (31) in form (35):

$$-(K+k) \cdot x + K \cdot y - k \cdot x_0 - m_S^* \cdot \omega^2 \cdot x^{II} = \frac{m_T^* \cdot \omega^2 \cdot y^{II} \cdot y^{I}}{x^{I}} \qquad (35)$$

The equation (35), can be amplified by x' and obtains the relation (36):

$$-(K+k) \cdot x \cdot x^{I} + K \cdot y \cdot x^{I} - k \cdot x_0 \cdot x^{I} -$$
$$- m_S^* \cdot \omega^2 \cdot x^{I} \cdot x^{II} = m_T^* \cdot \omega^2 \cdot y^{I} \cdot y^{II} \qquad (36)$$

Now, it replaces the term K.y.x' with $K \cdot y \cdot \frac{K}{K+k} \cdot y^{I}$, (taken in calculation the statically assumption, $F_m = F_r$) and it obtains the form (37):

$$-(K+k) \cdot x \cdot x^{I} + \frac{K^2}{K+k} \cdot y \cdot y^{I} - k \cdot x_0 \cdot x^{I} -$$
$$- m_S^* \cdot \omega^2 \cdot x^{I} \cdot x^{II} = m_T^* \cdot \omega^2 \cdot y^{I} \cdot y^{II} \qquad (37)$$

It integrates directly the equation (37) and obtains the mother equation (38):

$$-(K+k)\cdot\frac{x^2}{2}+\frac{K^2}{K+k}\cdot\frac{y^2}{2}-k\cdot x_0\cdot x- \tag{38}$$
$$-m_S^*\cdot\omega^2\cdot\frac{x'^2}{2}=m_T^*\cdot\omega^2\cdot\frac{y'^2}{2}+C$$

With the initial condition, at the φ=0, y=y'=0 and x=x'=0, it obtains for the constant of integration, C the value 0. In this case the equation (38), takes the form (39):

$$-(K+k)\cdot\frac{x^2}{2}+\frac{K^2}{K+k}\cdot\frac{y^2}{2}-k\cdot x_0\cdot x-m_S^*\cdot\omega^2\cdot\frac{x'^2}{2}=m_T^*\cdot\omega^2\cdot\frac{y'^2}{2} \tag{39}$$

The equation (39) can be put in the form (40), if one divides it with the $-\frac{K+k}{2}$:

$$x^2+2\cdot\frac{k\cdot x_0}{K+k}\cdot x+\frac{m_S^*\cdot\omega^2}{K+k}\cdot x'^2+\frac{m_T^*\cdot\omega^2}{K+k}y'^2-\frac{K^2}{(K+k)^2}\cdot y^2=0 \tag{40}$$

The mother equation (40), take the form (41), if one notes: $x'=\frac{K}{K+k}\cdot y'$, (the static assumption, $F_m=F_r$).

$$x^2+2\cdot\frac{k\cdot x_0}{K+k}\cdot x-\frac{K^2}{(K+k)^2}\cdot y^2+\frac{\frac{K^2}{(K+k)^2}\cdot m_S^*+m_T^*}{(K+k)}\cdot\omega^2\cdot y'^2=0 \tag{41}$$

3.1. Solving the mother equation (41) directly

The equation (41) is a two degree equation in x; One determines directly, Δ (42-43) and $X_{1,2}$ (44):

$$\Delta=\frac{(k\cdot x_0)^2+(K\cdot s)^2}{(K+k)^2}-\frac{m_S^*\cdot\frac{K^2}{(K+k)^2}+m_T^*}{(K+k)}\cdot y'^2\cdot\omega^2 \tag{42}$$

$$\Delta=\frac{(k\cdot x_0)^2+(K\cdot s)^2}{(K+k)^2}-\frac{m_S^*\cdot\frac{K^2}{(K+k)^2}+m_T^*}{(K+k)}\cdot(D\cdot s')^2\cdot\omega^2 \tag{43}$$

$$X_{1,2}=-\frac{k\cdot x_0}{K+k}\pm\sqrt{\Delta} \tag{44}$$

Physically, just the positive solution is valid (see the relation 45):

$$X=\sqrt{\Delta}-\frac{k\cdot x_0}{K+k} \tag{45}$$

3.2. Solving the mother equation (41) with finished differences

We can solve the mother equation (41) using the finished differences. We notes:

$$X=s+\Delta X \tag{46}$$

With the notation (46) placed in the mother equation (41), it obtains the equation (47):

$$s^2 + (\Delta X)^2 + 2 \cdot \Delta X \cdot s + 2 \cdot \frac{k \cdot x_0}{K+k} \cdot s + 2 \cdot \frac{k \cdot x_0}{K+k} \cdot \Delta X - \frac{K^2}{(K+k)^2} s^2 + \frac{\frac{K^2}{(K+k)^2} \cdot m_S^* + m_T^*}{(K+k)} \cdot \omega^2 \cdot y'^2 = 0 \qquad (47)$$

The equation (47) is a two degree equation in ΔX, which can be solved directly with Δ (49) and $\Delta X_{1,2}$, (50), or transformed in a single degree equation in ΔX, with $(\Delta X)^2 \cong 0$, solved by the relation (48).

$$\Delta X = (-1) \cdot \frac{(k^2 + 2 \cdot k \cdot K) \cdot s^2 + 2 \cdot k \cdot x_0 \cdot (K+k) \cdot s + [\frac{K^2}{K+k} \cdot m_S^* + (K+k) \cdot m_T^*] \cdot \omega^2 \cdot (Ds')^2}{2 \cdot (s + \frac{k \cdot x_0}{K+k}) \cdot (K+k)^2} \qquad (48)$$

$$\Delta = \frac{K^2 \cdot s^2 + k^2 \cdot x_0^2 - [\frac{K^2}{K+k} \cdot m_S^* + (K+k) \cdot m_T^*] \cdot \omega^2 \cdot (D \cdot s')^2}{(K+k)^2} \qquad (49)$$

$$\Delta X = \sqrt{\Delta} - (s + \frac{k \cdot x_0}{K+k}) \qquad (50)$$

CONCLUSION

The direct integration of the differential equation (31) generates the mother equation (41), which can be solved directly, with the relation (48). "D" represents the dynamic transmission function (the dynamic transmission coefficient).

REFERENCES

[1] Antonescu, P., Oprean, M., Petrescu, Fl., *Analiza dinamică a mecanismelor de distribuţie cu came,* In: The Proceedings of 7th National Symposium on RIMS, MERO'87, Bucureşti, vol. 3, pp. 126-133, 1987.
[2] Antonescu, P., Petrescu, Fl., *Contributii la analiza cinetoelastodinamică a mecanismelor de distribuţie,* In: The Proceedings of 5th International Symposium on TMM, SYROM'89, Bucureşti, pp. 33-40, 1989.
[3] Petrescu, F., Petrescu, R., *Elemente de dinamica mecanismelor cu came,* In: The Proceedings of 7th National Symposium, PRASIC'02, Braşov, vol. I, pp. 327-332, 2002.

CHAPTER VII
CAM GEARS DYNAMICS ILLUSTRATED IN THE CLASSIC DISTRIBUTION

Abstract: *The chapter presents an original method to determine the general dynamics of mechanisms with rotation cams and followers, particularized to the plate translated follower. First, it presents the dynamics kinematics. Then it solves the Lagrange equation and using an original dynamic model with one degree of freedom, with variable internal amortization, it makes the dynamic analysis.*
Keywords: *cam dynamics, classic distribution, cams, followers, dynamics*

1 Introduction

The chapter proposes an original dynamic model illustrated for the rotating cam with plate translate follower. It presents the **dynamics kinematics** (the original kinematics); the variable velocity of the camshaft obtained by an approximate method is used with an original dynamic system having one degree of freedom and a variable internal amortization [1]; it tests two movement laws, one classic and the other original.

2 Dynamics of the classic distribution mechanism
2.1 Precision kinematics in the classic distribution mechanism

In the picture number one, it presents the kinematic schema of the classic distribution mechanism, in two consecutive positions; with an interrupted line is represented the particular position when the follower is situated in the lowest possible plane, (s=0), and the cam which has a clockwise rotation, with constant angular velocity, ω, is situated in the point A^0, (the fillet point between the base profile and the rise profile), a particular point that marks the beginning of the rise movement of the follower, imposed by the cam-profile; with a continue line is represented the higher joint in a certain position of the rise phase.

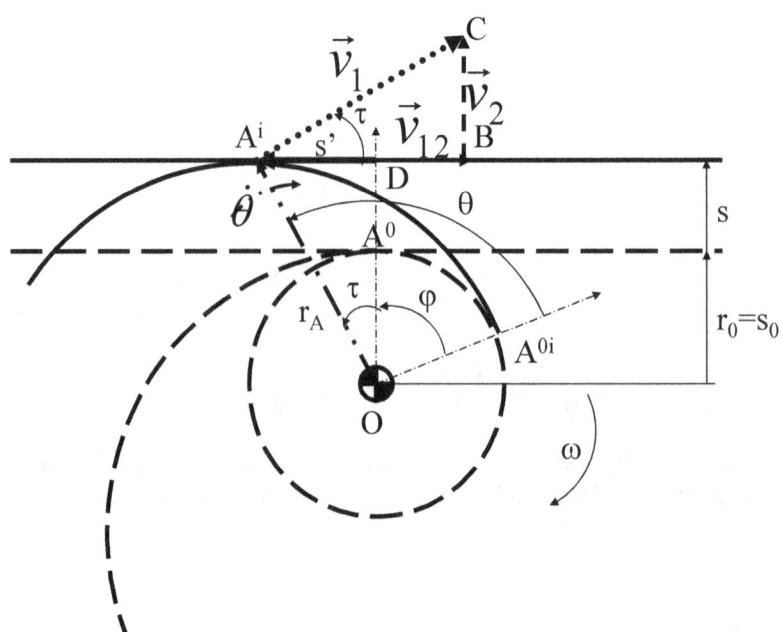

Fig. 1 *The kinematics of the classic distribution mechanism*

The point A^0, which marks the initial higher pair, represents in the same time the contact point between the cam and the follower in the first position. The cam is rotating with the angular velocity, ω (the camshaft angular velocity), describing the angle φ, which shows how the base circle has rotated clockwise (together with the camshaft); this rotation can be seen on the base circle between the two particular points, A^0 and A^{0i}.

In this time the vector r_A=OA (which represents the distance between the centre of cam O, and the contact point A), has rotated anticlockwise with the angle τ. If one measures the angle θ, which positions the general vector, r_A, in function of the particular vector, r_{A0}, it obtains the relation (0):

$$\theta = \varphi + \tau \tag{0}$$

where r_A is the module of the vector $\vec{r}_A$, and θ_A represents the phase angle of the vector $\vec{r}_A$.

The angular velocity of the vector $\vec{r}_A$ is $\dot{\theta}_A$ which is a function of the angular velocity of the camshaft, ω, and of the angle φ (by the movement laws s(φ), s'(φ), s''(φ)).

The follower isn't acted directly by the angle φ and the angular velocity ω; it's acted by the vector $\vec{r}_A$, which has the module r_A, the position angle θ_A and the angular velocity $\dot{\theta}_A$. From here we deduce a particular (dynamic) kinematics, the classical kinematics being just static and approximate kinematics.

Kinematic, it defines the next velocities (Fig. 1).

$\vec{v}_1$ =the cam's velocity; which is the velocity of the vector $\vec{r}_A$, in the point A; now the classical relation (1) becomes an approximate relation, and the real relation takes the form (2).

$$v_1 = r_A \cdot \omega \tag{1}$$

$$v_1 = r_A \cdot \dot{\theta}_A \tag{2}$$

The velocity $\vec{v}_1 = AC$ is separating into the velocity $\vec{v}_2$=BC (the follower's velocity which acts on its axe, vertically) and $\vec{v}_{12}$=AB (the slide velocity between the two profiles, the sliding velocity between the cam and the follower, which works along the direction of the commune tangent line of the two profiles in the contact point).

Because usually the cam profile is synthesis for the classical module C with the AD=s' known, we can write the relations:

$$r_A^2 = (r_0 + s)^2 + s'^2 \tag{3}$$

$$r_A = \sqrt{(r_0 + s)^2 + s'^2} \tag{4}$$

$$\cos \tau = \frac{r_0 + s}{r_A} = \frac{r_0 + s}{\sqrt{(r_0 + s)^2 + s'^2}} \tag{5}$$

$$\sin \tau = \frac{AD}{r_A} = \frac{s'}{r_A} = \frac{s'}{\sqrt{(r_0 + s)^2 + s'^2}} \tag{6}$$

$$v_2 = v_1 \cdot \sin \tau = r_A \cdot \dot{\theta}_A \cdot \frac{s'}{r_A} = s' \cdot \dot{\theta}_A \tag{7}$$

Now, the follower's velocity isn't $\dot{s}$ ($v_2 \neq \dot{s} \equiv s' \cdot \omega$), but it's given by the relation (9). In the case of the classical distribution mechanism the transmitting function D is given by the relations (8):

47

$$\dot{\theta}_A = D.\omega$$
$$D = \frac{\dot{\theta}_A}{\omega} \qquad (8)$$

$$v_2 = s'.\dot{\theta}_A = s'.D.\omega \qquad (9)$$

The determining of the sliding velocity between the profiles is made with the relation (10):

$$v_{12} = v_1.\cos\tau = r_A.\dot{\theta}_A.\frac{r_0+s}{r_A} = (r_0+s).\dot{\theta}_A \qquad (10)$$

The angles τ and θ_A will be determined, and also their first and second derivatives.

The τ angle has been determined from the triangle ODA^i (Fig.1) with the relations (11-13):

$$\sin\tau = \frac{s'}{\sqrt{(r_0+s)^2+s'^2}} \qquad (11)$$

$$\cos\tau = \frac{r_0+s}{\sqrt{(r_0+s)^2+s'^2}} \qquad (12)$$

$$tg\tau = \frac{s'}{r_0+s} \qquad (13)$$

It derives (11) in function of φ angle and obtains (14):

$$\tau'.\cos\tau = \frac{s''.r_A - s'.\dfrac{(r_0+s).s'+s'.s''}{r_A}}{(r_0+s)^2+s'^2} \qquad (14)$$

The relation (14) will be written in the form (15):

$$\tau'.\cos\tau = \frac{s''.(r_0+s)^2 + s''.s'^2 - s'^2.(r_0+s) - s'^2.s''}{[(r_0+s)^2+s'^2].\sqrt{(r_0+s)^2+s'^2}} \qquad (15)$$

From the relation (12) it extracts the value of $\cos\tau$, which will be introduced in the left term of the expression (15); then we can reduce $s''.s'^2$ from the right term of the expression (15) and it obtains the relation (16):

$$\tau'.\frac{r_0+s}{\sqrt{(r_0+s)^2+s'^2}} = \frac{(r_0+s).[s''.(r_0+s)-s'^2]}{[(r_0+s)^2+s'^2].\sqrt{(r_0+s)^2+s'^2}} \qquad (16)$$

After some simplifications the relation (17), which represents the expression of τ', is finally obtained:

$$\tau' = \frac{s''.(r_0+s)-s'^2}{(r_0+s)^2+s'^2} \qquad (17)$$

Now when τ' has been explicitly deduced, the next derivatives can be determined. The expression (17) will be derived directly and it obtains for the beginning the relation (18):

$$\tau'' = \frac{[s'''(r_0+s)+s''s'-2s's''][(r_0+s)^2+s'^2]-2[s''(r_0+s)-s'^2][(r_0+s)s'+s's'']}{[(r_0+s)^2+s'^2]^2} \quad (18)$$

The terms from the first bracket of the numerator (s'.s'') are reduced, and then it draws out s' from the fourth bracket of the numerator and obtains the expression (19):

$$\tau'' = \frac{[s'''\cdot(r_0+s)-s'\cdot s''].[(r_0+s)^2+s'^2]-2\cdot s'\cdot[s''\cdot(r_0+s)-s'^2]\cdot[r_0+s+s'']}{[(r_0+s)^2+s'^2]^2} \quad (19)$$

Now we can calculate θ_A, with its first two derivatives, $\dot{\theta}_A$ and $\ddot{\theta}_A$. We will write θ instead of θ_A, to simplify the notation. It determines the relation (20) which is the same of (0):

$$\theta = \tau + \varphi \quad (20)$$

We derive the relation (20) and one obtains the expression (21):

$$\dot{\theta} = \dot{\tau} + \dot{\varphi} = \tau'\cdot\omega + \omega = \omega\cdot(1+\tau') = D\cdot\omega \quad (21)$$

It derives twice (20), or derives (21) and obtains (22):

$$\ddot{\theta} = \ddot{\tau} + \ddot{\varphi} = \tau''\cdot\omega^2 = D'\cdot\omega^2 \quad (22)$$

We can write now the transmission functions, D and D' (for the classical module, C), in the forms (23-24):

$$D = \tau'+1 \quad (23)$$

$$D^I = \tau'' \quad (24)$$

To calculate the follower's velocity (25) we need the expression of the transmission function, D.

$$v_2 = s'\cdot w = s'\cdot\dot{\theta}_A = s'\cdot\dot{\theta} = s'\cdot D\cdot\omega = \dot{s}\cdot D \quad (25)$$

Where:

$$w = D\cdot\omega \quad (26)$$

For the classical distribution mechanism (Module C), the variable w is the same as $\dot{\theta}_A$ (see the relation 25).

But in the case of B and F modules (at the cam gears where the follower has a roll), the transmitted function D and w take complex forms.

We can determine now the acceleration of the follower (27).

$$\ddot{y} \equiv a_2 = (s''\cdot D + s'\cdot D')\cdot\omega^2 \quad (27)$$

Figure 2 represents the classical and dynamic kinematics; the velocities (a), and the accelerations (b).

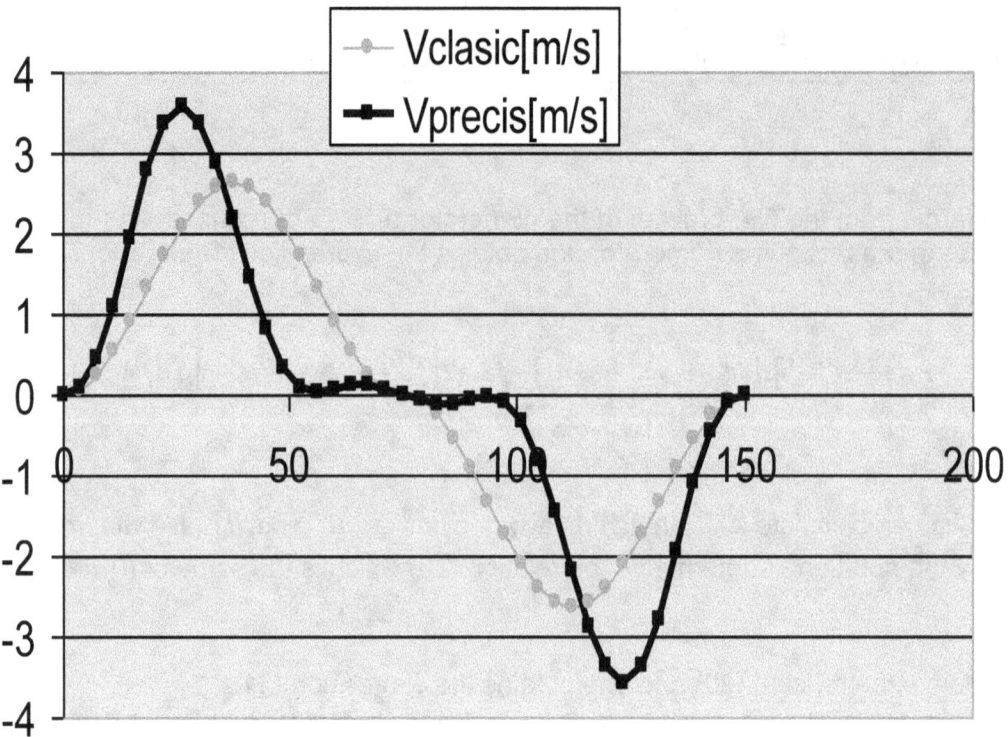

Fig. 2a *The classical and dynamic kinematics; velocities of the follower*

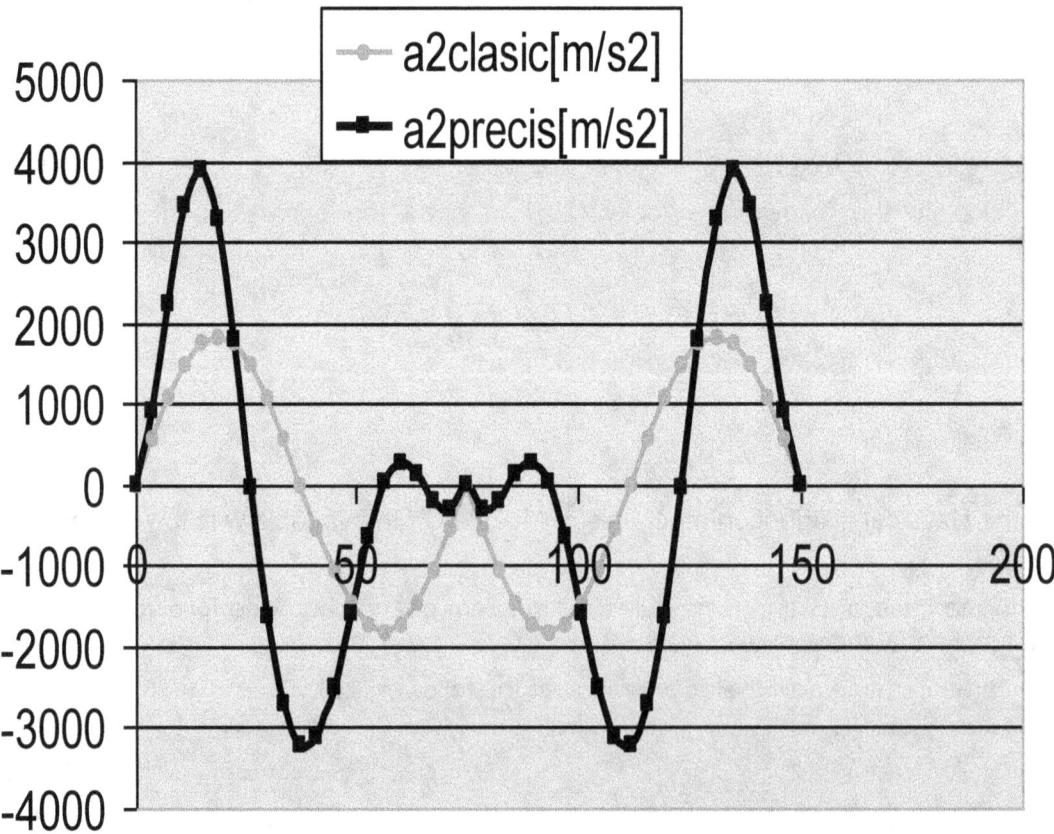

Fig. 2b *The classical and dynamic kinematics; accelerations of the follower*

To determine the acceleration of the follower, s' and s", D and D', τ' and τ" are necessary be known.

The dynamic kinematics diagrams of v_2 (obtained with relation 25, see Fig. 2a), and a_2 (obtained with relation 27, see Fig. 2b), have a more dynamic aspect than one kinematic (classic).

It has used the movement law SIN, a rotational speed of the crankshaft n=5500 rpm, a rise angle $\varphi_u=75^0$, a fall angle $\varphi_d=75^0$ (identically with the ascendant angle), a ray of the basic circle of the cam, $r_0=17$ mm and a maxim stroke of the follower, $h_T=6$ mm.

Anyway, the dynamics is more complex, having in view the masses and the inertia moments, the resistant and motor forces, the elasticity constants and the amortization coefficient of the kinematic chain, the inertia forces of the system, the angular velocity of the camshaft and the variation of the camshaft's angular velocity, ω, with the cam's position, φ, and with the rotational speed of the crankshaft, n.

2.2 Solving approximately the Lagrange movement equation

In the kinematics and the static forces study of the mechanisms one considers the shaft's angular velocity constant, $\dot{\varphi}=\omega=$constant, and the angular acceleration null, $\ddot{\varphi}=\dot{\omega}=\varepsilon=0$. In reality, this angular velocity ω isn't constant, it is variable with the camshaft position, φ.

In mechanisms with cam and follower the camshaft's angular velocity is variable as well. We shall see further the Lagrange equation, written in the differentiate mode and its general solution. The differentiate Lagrange equation has the form (28):

$$J^*.\ddot{\varphi}+\frac{1}{2}.J^{*I}.\dot{\varphi}^2 = M^* \qquad (28)$$

Where J* is the mechanical inertia moment (mass moment, or mechanic moment) of the mechanism, reduced at the crank, and M* represents the difference between the motor moment reduced at the crank and the resistant moment reduced at the crank; the angle φ represents the rotation angle of the crank (crankshaft). J^{*I} represents the derivative of the mechanic moment in function of the rotation angle φ of the crank (29).

$$\frac{1}{2}.J^{*I} = \frac{1}{2}.\frac{dJ^*}{d\varphi} = L \qquad (29)$$

Using the notation (29), the equation (28) will be written in the form (30):

$$J^*.\ddot{\varphi}+L.\dot{\varphi}^2 = M^* \qquad (30)$$

We divide the terms by J* and (30) takes the form (31):

$$\ddot{\varphi}+\frac{L}{J^*}.\dot{\varphi}^2 = \frac{M^*}{J^*} \qquad (31)$$

The term with $\dot{\varphi}^2$ will be moved to the right side of the equation and the form (32) will be obtained:

$$\ddot{\varphi} = \frac{M^*}{J^*} - \frac{L}{J^*}.\dot{\varphi}^2 \qquad (32)$$

Replacing the left term of the expression (32) with (33) we obtain the relation (34):

$$\ddot{\varphi} = \frac{d\dot{\varphi}}{dt} = \frac{d\dot{\varphi}}{d\varphi} \cdot \frac{d\varphi}{dt} = \frac{d\dot{\varphi}}{d\varphi} \cdot \dot{\varphi} = \frac{d\omega}{d\varphi} \cdot \omega \tag{33}$$

$$\omega \cdot \frac{d\omega}{d\varphi} = \frac{M^*}{J^*} - \frac{L}{J^*} \cdot \omega^2 = \frac{M^* - L \cdot \omega^2}{J^*} \tag{34}$$

Because, for an angle φ, ω is different from the nominal constant value ω_n, it can write the relation (35), where dω represents the momentary variation for the angle φ; the variable dω and the constant ω_n lead us to the needed variable, ω:

$$\omega = \omega_n + d\omega \tag{35}$$

In the relation (35), ω and dω are functions of the angle φ, and ω_n is a constant parameter, which can take different values in function of the rotational speed of the drive-shaft, n. At a moment, n is a constant and ω_n is a constant as well (because ω_n is a function of n). The angular velocity, ω, becomes a function of n too (see the relation 36):

$$\omega(\varphi, n) = \omega_n(n) + d\omega(\varphi, \omega_n(n)) \tag{36}$$

With (35) in (34), it obtains the equation (37):

$$(\omega_n + d\omega) \cdot d\omega = [\frac{M^*}{J^*} - \frac{L}{J^*} \cdot (\omega_n + d\omega)^2] \cdot d\varphi \tag{37}$$

The relation (37) takes the form (38):

$$\omega_n \cdot d\omega + (d\omega)^2 = \frac{M^*}{J^*} \cdot d\varphi - \frac{L}{J^*} \cdot d\varphi \cdot [\omega_n^2 + (d\omega)^2 + 2 \cdot \omega_n \cdot d\omega] \tag{38}$$

The equation (38) will be written in the form (39):

$$\omega_n \cdot d\omega + (d\omega)^2 - \frac{M^*}{J^*} \cdot d\varphi + \frac{L}{J^*} \cdot d\varphi \cdot \omega_n^2 + \\ + \frac{L}{J^*} \cdot d\varphi \cdot (d\omega)^2 + 2 \cdot \frac{L}{J^*} \cdot d\varphi \cdot \omega_n \cdot d\omega = 0 \tag{39}$$

The relation (39) takes the form (40):

$$(\frac{L}{J^*} \cdot d\varphi + 1) \cdot (d\omega)^2 + 2 \cdot (\frac{L}{J^*} \cdot d\varphi + \frac{1}{2}) \cdot \omega_n \cdot d\omega - \\ - (\frac{M^*}{J^*} \cdot d\varphi - \frac{L}{J^*} \cdot d\varphi \cdot \omega_n^2) = 0 \tag{40}$$

The relation (40) is an equation of the second degree in dω. The discriminate of the equation (40) can be written in the forms (41) and (42):

$$\Delta = \frac{L^2}{J^{*2}} \cdot (d\varphi)^2 \cdot \omega_n^2 + \frac{\omega_n^2}{4} + \frac{L}{J^*} \cdot d\varphi \cdot \omega_n^2 + \frac{L \cdot M^*}{J^{*2}} \cdot (d\varphi)^2 \\ + \frac{M^*}{J^*} \cdot d\varphi - \frac{L^2}{J^{*2}} \cdot (d\varphi)^2 \cdot \omega_n^2 - \frac{L}{J^*} \cdot d\varphi \cdot \omega_n^2 \tag{41}$$

$$\Delta = \frac{\omega_n^2}{4} + \frac{L \cdot M^*}{J^{*2}} \cdot (d\varphi)^2 + \frac{M^*}{J^*} \cdot d\varphi \tag{42}$$

We keep for dω just the positive solution, which can generate positives and negatives normal values (43), and in this mode only normal values will be obtained for ω; for $\Delta<0$ it considers dω=0 (this case must be not seeing if the equation is correct).

$$d\omega = \frac{-\frac{L}{J^*}.d\varphi.\omega_n - \frac{\omega_n}{2} + \sqrt{\Delta}}{\frac{L}{J^*}.d\varphi + 1} \qquad (43)$$

Observations: For mechanisms with rotate cam and follower, using the new relations, with M^* (the reduced moment of the mechanism) obtained by the writing of the known reduced resistant moment and by the determination of the reduced motor moment by the integration of the resistant moment it frequently obtains some bigger values for dω, or zones with Δ negative, with complex solutions for dω. This fact gives us the obligation to reconsider the method to determine the reduced moment.

If we take into consideration M^*_r and M^*_m, calculated independently (without integration), it obtains for the mechanisms with cam and follower normal values for dω, and $\Delta \geq 0$.

In paper [1] it presents the relations to determine the resistant force (44) reduced to the valve, and the motor force (45) reduced to the ax of the valve:

$$F_r^* = k.(x_0 + x) \qquad (44)$$

$$F_m^* = K.(y - x) \qquad (45)$$

The reduced resistant moment (46), or the reduced motor moment (47), can be obtained by the resistant or motor force multiplied by the reduced velocity, x'.

$$M_r^* = k.(x_0 + x).x' \qquad (46)$$

$$M_m^* = K.(y - x).x' \qquad (47)$$

2.3 The dynamic relations used

The dynamics relations used (48-49), have been deduced in the paper [1]:

$$\Delta X = (-1) \cdot \frac{(k^2 + 2 \cdot k \cdot K) \cdot s^2 + 2 \cdot k \cdot x_0 \cdot (K+k) \cdot s + [\frac{K^2}{K+k} \cdot m_S^* + (K+k) \cdot m_T^*] \cdot \omega^2 \cdot (Ds')^2}{2 \cdot (s + \frac{k \cdot x_0}{K+k}) \cdot (K+k)^2} \qquad (48)$$

$$X = s - \frac{[\frac{K^2}{K+k} \cdot m_S^* + (K+k) \cdot m_T^*] \cdot \omega^2 \cdot (Ds')^2}{2 \cdot (s + \frac{k \cdot x_0}{K+k}) \cdot (K+k)^2} - \frac{(k^2 + 2 \cdot k \cdot K) \cdot s^2 + 2 \cdot k \cdot x_0 \cdot (K+k) \cdot s}{2 \cdot (s + \frac{k \cdot x_0}{K+k}) \cdot (K+k)^2} \qquad (49)$$

2.4 The dynamic analysis

The dynamic analysis or the classical movement law sin, can be seen in the diagram from figure 3, and in figure 4 one can see the diagram of an original movement law (C4P) (module C).

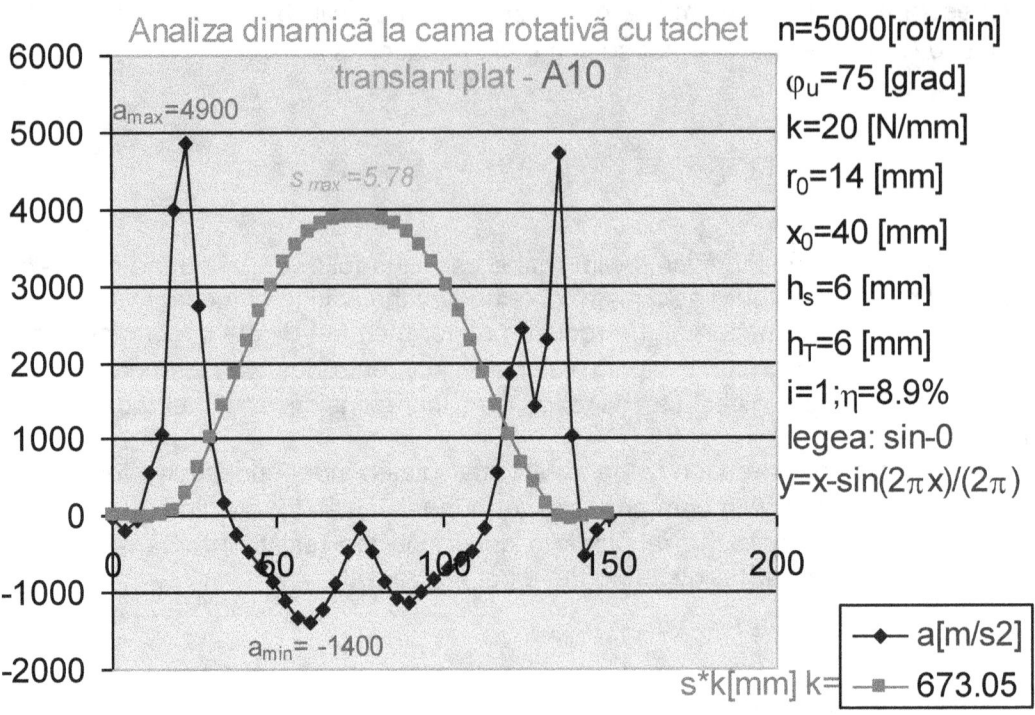

Fig. 3 *The dynamic analysis of the law sin, Module C, $\varphi_u=75^0$, n=5000 rpm*

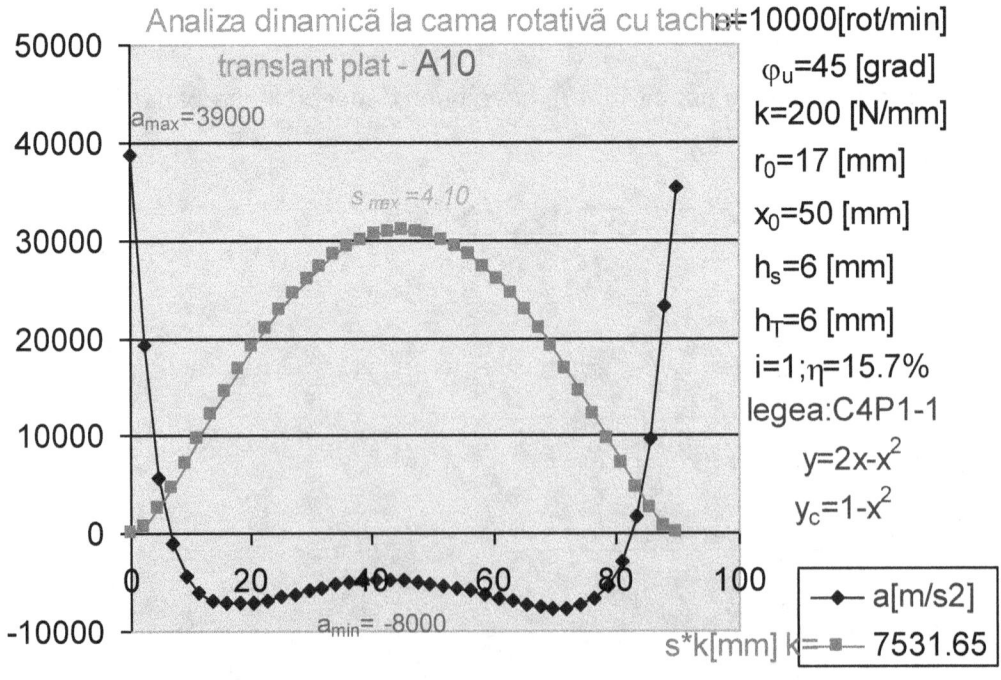

Fig. 4 *The dynamic analysis of the new law, C4P, Module C, $\varphi_u=45^0$, n=10000 rpm*

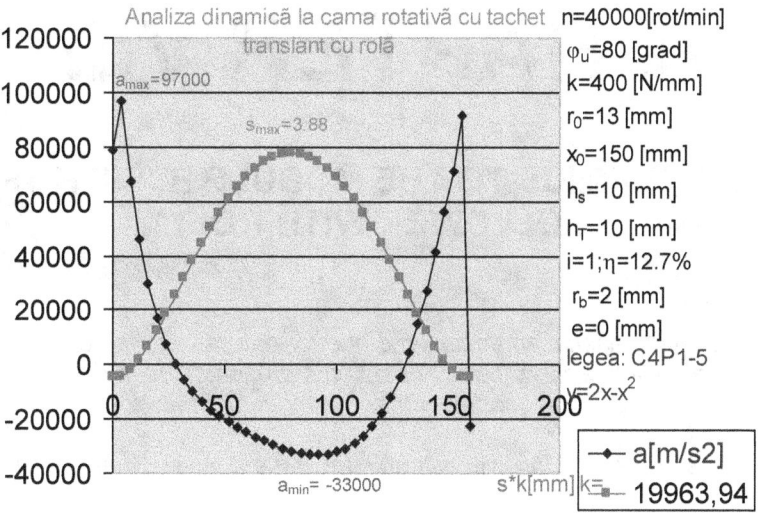

Fig. 5 Law C4P1-5, Module B, $\varphi_u=80^0$, $n=40000$ rpm

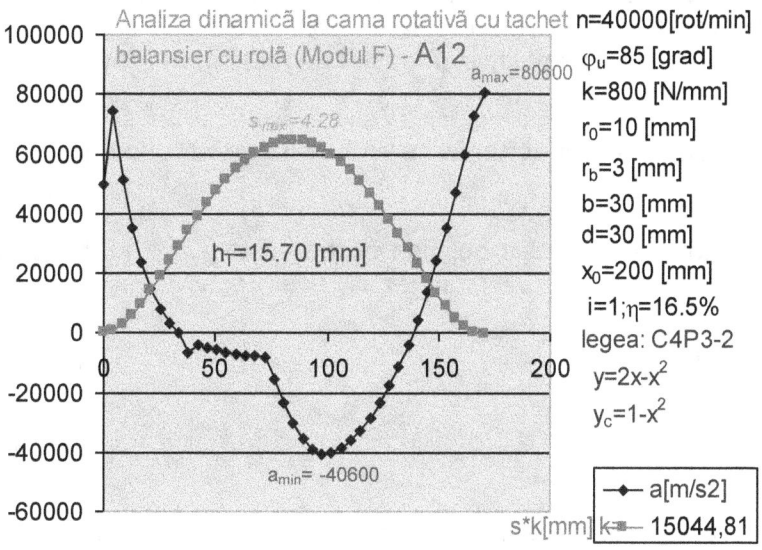

Fig. 6 Law C4P3-2, Module F, $\varphi_u=85^0$, $n=40000$ rpm

3 Conclusions

Using the classical movement laws, the dynamics of the distribution cam-gears depreciate rapidly at the increasing of the rotational speed of the shaft. To support a high rotational speed it is necessary the synthesis of the cam-profile by new movement laws, and for the new Modules.

A new and original movement law is presented in the pictures number 4, 5 and 6; it allows the increase of the rotational speed to the values: 10000-20000 rpm, in the classical module C presented (Fig. 4). With others modules (B, F) it can obtain 30000-40000 rpm (see Figs. 5, 6).

References

[1] Petrescu F.I., Petrescu R.V., *Contributions at the dynamics of cams*. In the Ninth IFToMM International Sympozium on Theory of Machines and Mechanisms, SYROM 2005, Bucharest, Romania, Vol. I, pp. 123-128, 2005.

CHAPTER VIII

CAM GEARS DYNAMICS TO THE MODULE B (WITH TRANSLATED FOLLOWER WITH ROLL)

Abstract: The chapter briefly presents an original method for determining the dynamics of mechanisms with rotation cam and translated follower with roll. First, one presents the dynamics kinematics. Then one performs the dynamic analysis of a few models, for some movement laws, imposed on the follower, by the designed cam profile.
Keywords: cam dynamics, translated follower with roll, movement laws, dynamics kinematic

1 Introduction

The chapter proposes an original dynamic model of the cam gear with a translated follower with a roll. First, one presents the *dynamics kinematics*. Then one performs the dynamic analysis of a few models, for some movement laws, imposed on the follower, by the designed cam profile.

2 The dynamics of distribution mechanisms with translated follower with roll
2.1 Generalities

The angle α_0 defines the basic position of the vector, $\bar{r}_{B0}$, in the OCB_0 triangle having a right angle (1-4):

$$r_{B_0} = r_0 + r_b \quad (1) \qquad\qquad s_0 = \sqrt{r_{B_0}^2 - e^2} \quad (2)$$

$$\cos\alpha_0 = \frac{e}{r_{B_0}} \quad (3) \qquad\qquad \sin\alpha_0 = \frac{s_0}{r_{B_0}} \quad (4)$$

The pressure angle, δ, between the normal n (which passes through the contact point A) and a vertical line, can be calculated with relations (5-7).

$$\cos\delta = \frac{s_0 + s}{\sqrt{(s_0 + s)^2 + (s'-e)^2}} \tag{5}$$

$$\sin\delta = \frac{s'-e}{\sqrt{(s_0 + s)^2 + (s'-e)^2}} \tag{6}$$

$$tg\delta = \frac{s'-e}{s_0 + s} \tag{7}$$

The vector $\bar{r}_A$ can be determined with relations (8-9):

$$r_A^2 = (e + r_b \cdot \sin\delta)^2 + (s_0 + s - r_b \cdot \cos\delta)^2 \tag{8}$$

$$r_A = \sqrt{(e + r_b \cdot \sin\delta)^2 + (s_0 + s - r_b \cdot \cos\delta)^2} \tag{9}$$

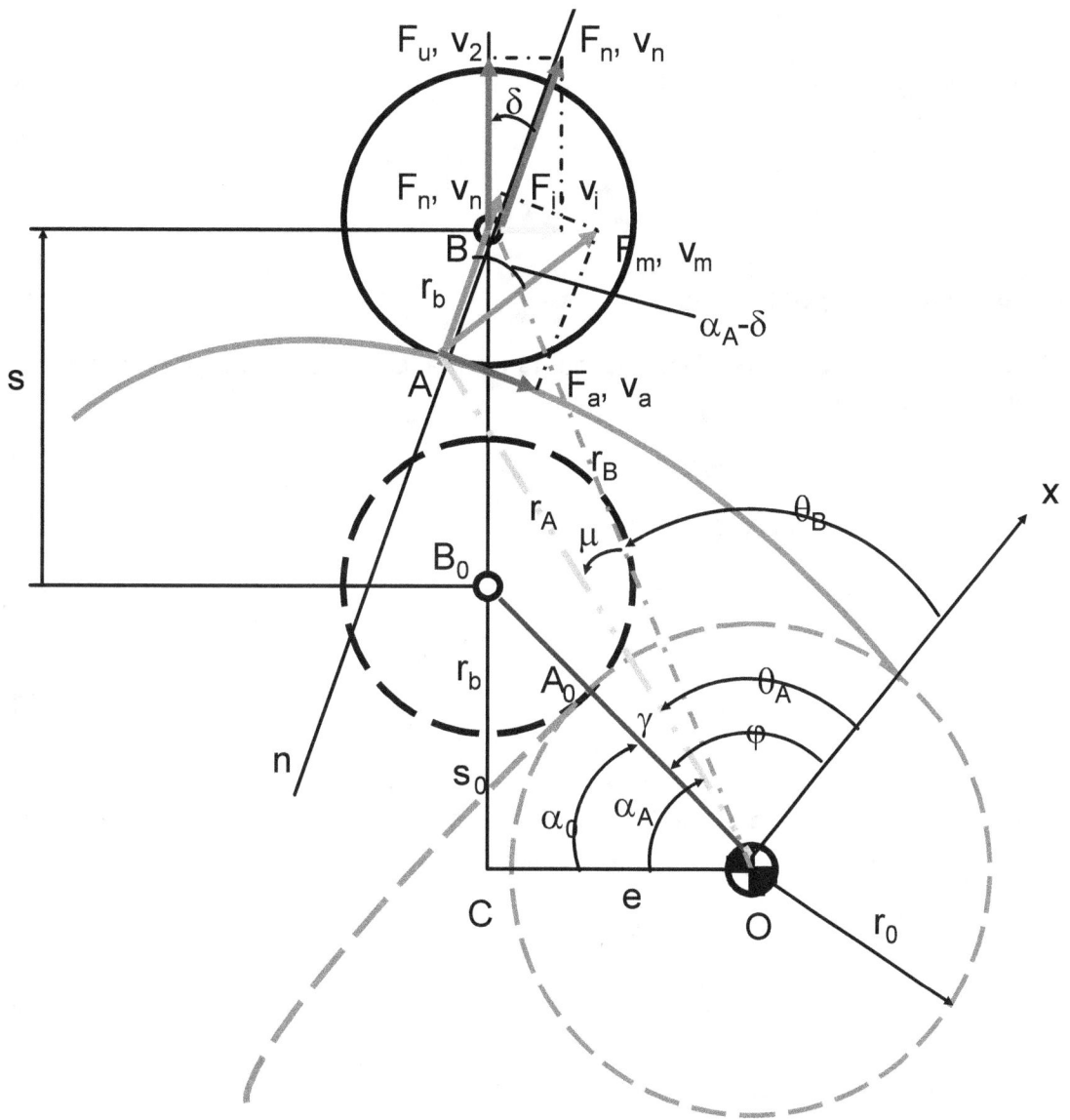

Fig. 1 *Mechanism with rotating cam and translating follower with roll*

We can calculate α_A (10-11):

$$\cos\alpha_A = \frac{e + r_b \cdot \sin\delta}{r_A} \qquad (10)$$

$$\sin\alpha_A = \frac{s_0 + s - r_b \cdot \cos\delta}{r_A} \qquad (11)$$

2.2 The relations to design the profile

$$\gamma = \alpha_A - \alpha_0 \qquad (12)$$

$$\cos\gamma - \cos\alpha_A \cdot \cos\alpha_0 + \sin\alpha_A \cdot \sin\alpha_0 \qquad (13)$$

$$\sin\gamma = \sin\alpha_A \cdot \cos\alpha_0 - \cos\alpha_A \cdot \sin\alpha_0 \qquad (14)$$

$$\theta_A = \varphi - \gamma \qquad (15)$$

$$\cos\theta_A = \cos\varphi \cdot \cos\gamma + \sin\varphi \cdot \sin\gamma \qquad (16)$$

$$\sin\theta_A = \sin\varphi \cdot \cos\gamma - \sin\gamma \cdot \cos\varphi \qquad (17)$$

2.3 The exact kinematics of B Module

From the triangle OCB (fig. 1) the length r_B (OB) and the complementary angles α_B (COB) and τ (CBO) are determined.

$$r_B^2 = e^2 + (s_0 + s)^2 \qquad (18)$$

$$r_B = \sqrt{r_B^2} \qquad (19)$$

$$\cos\alpha_B \equiv \sin\tau = \frac{e}{r_B} \qquad (20)$$

$$\sin\alpha_B \equiv \cos\tau = \frac{s_0 + s}{r_B} \qquad (21)$$

From the general triangle OAB, where one knows OB, AB, and the angle between them, B (ABO, which is the sum of τ and δ), the length OA and the angle μ (AOB) can be determined:

$$\cos(\delta + \tau) = \cos\delta \cdot \cos\tau - \sin\delta \cdot \sin\tau \qquad (22)$$

$$r_A^2 = r_B^2 + r_b^2 - 2 \cdot r_b \cdot r_B \cdot \cos(\delta + \tau) \qquad (23)$$

$$\cos\mu = \frac{r_A^2 + r_B^2 - r_b^2}{2 \cdot r_A \cdot r_B} \qquad (24)$$

$$\sin(\delta + \tau) = \sin\delta \cdot \cos\tau + \sin\tau \cdot \cos\delta \qquad (25)$$

$$\sin\mu = \frac{r_b}{r_A} \cdot \sin(\delta + \tau) \qquad (26)$$

With α_B and μ we can deduce now α_A and $\dot\alpha_A$:

$$\alpha_A = \alpha_B - \mu \qquad (27)$$

$$\dot\alpha_A = \dot\alpha_B - \dot\mu \qquad (28)$$

From (20) one obtains $\dot\alpha_B$ (32), (see 29-32) where $\dot r_B$ (31) can be deduced from (18). Then, (33) will be obtained from (24):

$$-\sin\alpha_B \cdot \dot\alpha_B = -\frac{e \cdot \dot r_B}{r_B^2} \qquad (29)$$

$$\dot\alpha_B = \frac{e \cdot r_B \cdot \dot r_B}{(s_0 + s) \cdot r_B^2} \qquad (30)$$

$$2 \cdot r_B \cdot \dot r_B = 2 \cdot (s_0 + s) \cdot \dot s \qquad r_B \cdot \dot r_B = (s_0 + s) \cdot \dot s \qquad (31)$$

$$\dot\alpha_B = \frac{e \cdot (s_0 + s) \cdot \dot s}{(s_0 + s) \cdot r_B^2} = \frac{e \cdot \dot s}{r_B^2} \qquad (32)$$

$$2 \cdot \dot{r}_A \cdot r_B \cdot \cos\mu + 2 \cdot r_A \cdot \dot{r}_B \cdot \cos\mu - 2 \cdot r_A \cdot r_B \cdot \sin\mu \cdot \dot{\mu} = 2 \cdot r_A \cdot \dot{r}_A + 2 \cdot r_B \cdot \dot{r}_B \tag{33}$$

From (33) one writes $\dot{\mu}$ (38), but it is necessary to obtain first $\dot{r}_A$ (34) from expression (23):

$$2 \cdot r_A \cdot \dot{r}_A = 2 \cdot r_B \cdot \dot{r}_B - 2 \cdot r_b \cdot \dot{r}_B \cdot \cos(\delta + \tau) + 2 \cdot r_b \cdot r_B \cdot \sin(\delta + \tau) \cdot (\dot{\delta} + \dot{\tau}) \tag{34}$$

To solve (34) we need the derivatives $\dot{\delta}$ and $\dot{\tau}$. From (7) relations (35 and 36) will be obtained. $\dot{\tau}$ takes the form (37):

$$\delta' = \frac{s'' \cdot (s_0 + e) - s' \cdot (s' - e)}{(s_0 + s)^2 + (s' - e)^2} \tag{35} \qquad \dot{\delta} = \delta' \cdot \omega \tag{36}$$

$$\dot{\tau} = -\dot{\alpha}_B = -\frac{e \cdot \dot{s}}{r_B^2} \tag{37}$$

Now we can determine $\dot{\mu}$ (38), $\dot{\alpha}_A$ (28) and $\dot{\theta}_A$ (39):

$$\dot{\mu} = \frac{\dot{r}_A \cdot r_B \cdot \cos\mu + r_A \cdot \dot{r}_B \cdot \cos\mu - r_A \cdot \dot{r}_A - r_B \cdot \dot{r}_B}{r_A \cdot r_B \cdot \sin\mu} \tag{38}$$

$$\dot{\theta}_A = \dot{\varphi} - \dot{\gamma} = \omega - \dot{\alpha}_A \tag{39}$$

We write $\cos\alpha_A$ and $\sin\alpha_A$ (40-41):

$$\cos\alpha_A = \frac{e \cdot \sqrt{(s_0 + s)^2 + (s' - e)^2} + r_b \cdot (s' - e)}{r_A \cdot \sqrt{(s_0 + s)^2 + (s' - e)^2}} \tag{40}$$

$$\sin\alpha_A = \frac{(s_0 + s) \cdot [\sqrt{(s_0 + s)^2 + (s' - e)^2} - r_b]}{r_A \cdot \sqrt{(s_0 + s)^2 + (s' - e)^2}} \tag{41}$$

Further, we can obtain expression $\cos(\alpha_A - \delta)$ (42), and $\cos(\alpha_A - \delta) \cdot \cos\delta$ (43):

$$\cos(\alpha_A - \delta) = \frac{(s_0 + s) \cdot s'}{r_A \cdot \sqrt{(s_0 + s)^2 + (s' - e)^2}} = \frac{s'}{r_A} \cdot \cos\delta \tag{42}$$

$$\cos(\alpha_A - \delta) \cdot \cos\delta = \frac{s'}{r_A} \cdot \cos^2\delta \tag{43}$$

Finally the forces and the velocities are deduced as follows (48-50):

$$\begin{cases} v_a = v_m \cdot \sin(\alpha_A - \delta) \\ \\ F_a = F_m \cdot \sin(\alpha_A - \delta) \end{cases} \tag{44}$$

$$\begin{cases} v_n = v_m \cdot \cos(\alpha_A - \delta) \\ \\ F_n = F_m \cdot \cos(\alpha_A - \delta) \end{cases} \tag{45}$$

$$\begin{cases} v_i = v_n \cdot \sin\delta \\ F_i = F_n \cdot \sin\delta \end{cases} \quad (46)$$

$$\begin{cases} v_2 = v_n \cdot \cos\delta = v_m \cdot \cos(\alpha_A - \delta) \cdot \cos\delta \\ F_u = F_n \cdot \cos\delta = F_m \cdot \cos(\alpha_A - \delta) \cdot \cos\delta \end{cases} \quad (47)$$

2.4 Determining the efficiency of the Module B

$$P_u = F_u \cdot v_2 = F_m \cdot v_m \cdot \cos^2(\alpha_A - \delta) \cdot \cos^2\delta \quad (48)$$

$$P_c = F_m \cdot v_m \quad (49)$$

$$\eta_i = \frac{P_u}{P_c} = \frac{F_m \cdot v_m \cdot \cos^2(\alpha_A - \delta) \cdot \cos^2\delta}{F_m \cdot v_m} =$$
$$= \cos^2(\alpha_A - \delta) \cdot \cos^2\delta = [\cos(\alpha_A - \delta) \cdot \cos\delta]^2 = \quad (50)$$
$$-[\frac{s'}{r_A} \cdot \cos^2\delta]^2 = \frac{s'^2}{r_A^2} \cdot \cos^4\delta$$

2.5 Determining the transmission function D, for the Module B

The follower's velocity (47) can be written into the form (51):

$$v_2 = v_n \cdot \cos\delta = v_m \cdot \cos(\alpha_A - \delta) \cdot \cos\delta = v_m \cdot \frac{s'}{r_A} \cdot \cos^2\delta =$$
$$= r_A \cdot \dot{\theta}_A \cdot \frac{s'}{r_A} \cdot \cos^2\delta = \dot{\theta}_A \cdot s' \cdot \cos^2\delta = \theta_A^I \cdot \omega \cdot s' \cdot \cos^2\delta \quad (51)$$

With relations (51) and (52) we determine the transmission function (the dynamic modulus), D (53):

$$v_2 = s' \cdot D \cdot \omega \quad (52)$$

$$D = \theta_A^I \cdot \cos^2\delta \quad (53)$$

Expression $\cos^2\delta$ is known (54):

$$\cos^2\delta = \frac{(s_0 + s)^2}{(s_0 + s)^2 + (s' - e)^2} \quad (54)$$

The expression of the θ'_A is more difficult (55):

$$\theta_A^I = [(s_0+s)^2 + e^2 - e \cdot s' - r_b \cdot \sqrt{(s_0+s)^2 + (s'-e)^2}] \cdot$$
$$\{[(s_0+s)^2 + (s'-e)^2] \cdot \sqrt{(s_0+s)^2 + (s'-e)^2}$$
$$+ r_b \cdot [s'' \cdot (s_0+s) - s' \cdot (s'-e) - (s_0+s)^2 - (s'-e)^2]\} / \quad (55)$$
$$[(s_0+s)^2 + (s'-e)^2]/\{[(s_0+s)^2 + e^2 + r_b^2] \cdot$$
$$\cdot \sqrt{(s_0+s)^2 + (s'-e)^2} - 2 \cdot r_b \cdot [(s_0+s)^2 + e^2 - e \cdot s']\}$$

We will determine μ by its expressions (56-57):

$$\cos\mu = \frac{[(s_0+s)^2 + e^2] \cdot \sqrt{(s_0+s)^2 + (s'-e)^2} - r_b \cdot [(s_0+s)^2 + e^2 - e \cdot s']}{r_A \cdot r_B \cdot \sqrt{(s_0+s)^2 + (s'-e)^2}} \quad (56)$$

$$\sin\mu = \frac{r_b \cdot (s_0+s) \cdot s'}{r_A \cdot r_B \cdot \sqrt{(s_0+s)^2 + (s'-e)^2}} \quad (57)$$

2.6 The dynamics of the Module B

For the dynamics of the Module B the relations (58-60) are used:

$$\Delta X = -\frac{\frac{k^2+2kK}{(K+k)^2} \cdot s^2 + \frac{2kx_0}{K+k} \cdot s + \frac{[\frac{K^2}{(K+k)^2} \cdot m_S^* + m_T^*] \cdot \omega^2}{K+k} \cdot y'^2}{2 \cdot [s + \frac{kx_0}{K+k}]} \quad (58)$$

$$\Delta X = -\frac{\frac{k^2+2kK}{(K+k)^2} \cdot s^2 + \frac{2kx_0}{K+k} \cdot s + \frac{[\frac{K^2}{(K+k)^2} \cdot m_S^* + m_T^*] \cdot \omega^2}{K+k} \cdot (D \cdot s')^2}{2 \cdot [s + \frac{kx_0}{K+k}]} \quad (59)$$

$$X = s + \Delta X \quad (60)$$

2.7 The dynamic analysis of the module B

It presents now the dynamics of the module B for some known movement laws.

We begin with the classical law SIN (see the diagram in figure 2); A speed rotation n=5500 [rot/min], for a maxim theoretical displacement of the valve h=6 [mm] is used. The phase angle is $\varphi_u = \varphi_c = 65$ [degree]; the ray of the basic circle is $r_0 = 13$ [mm].

For the ray of the roll the value $r_b = 13$ [mm] has been adopted.

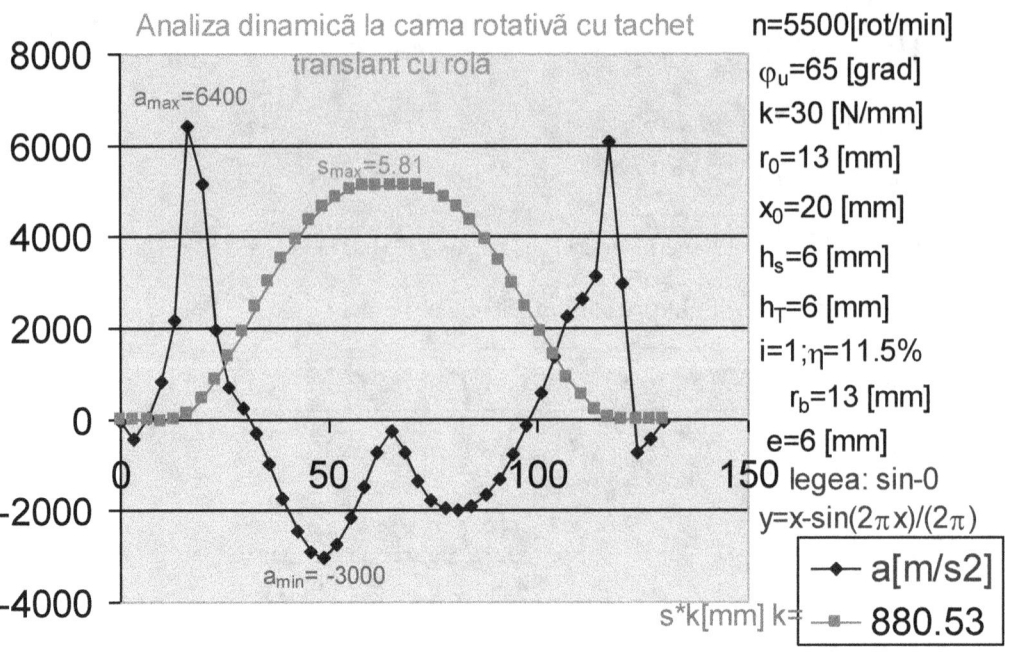

Fig. 2 *The dynamic analysis of the module B. The law SIN, n=550 rpm, $\varphi_u=65^0$, $r_0=13$ [mm], $r_b=13$ [mm], $h_T=6$ [mm], e=0 [mm], k=30 [N/mm], and $x_0=20$ [mm].*

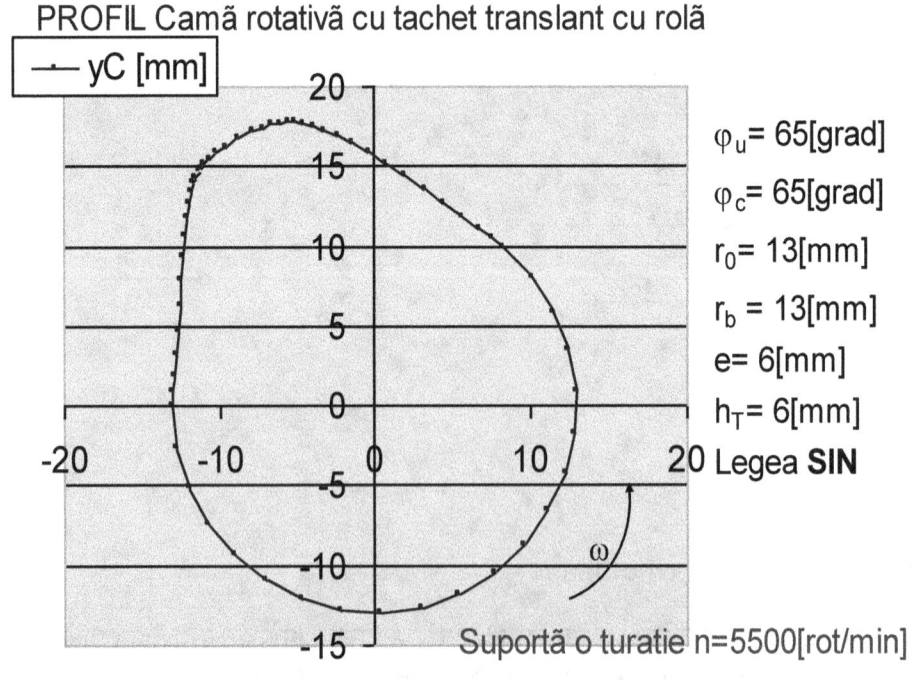

Fig. 3 *The profile SIN at the module B. n=5500 rpm $\varphi_u=65^0$, $r_0=13$ [mm], $r_b=13$ [mm], $h_T=6$ [mm].*

The dynamics are better than for the classical module C. *For a phase angle of just 65 degrees the accelerations have the same values as for the classical module C for a relaxed phase (75^0-80^0).*

In figure 3 we can see the cam's profile. It uses the profile sin, a rotation speed n=5500 rpm, and φ_u=65^0, r_0=13 [mm], r_b=13 [mm], h_T=6 [mm].

The law COS can be seen in figures 4 and 5.

In the figure 4 is presented the dynamic analyze of the profile cos, and its profile design can be seen in the figure 5.

The principal parameters are:

Law COS, n=5500 rpm, φ_u=65^0, r_0=13 [mm], r_b=6 [mm], h_T=6 [mm], η=10.5%.

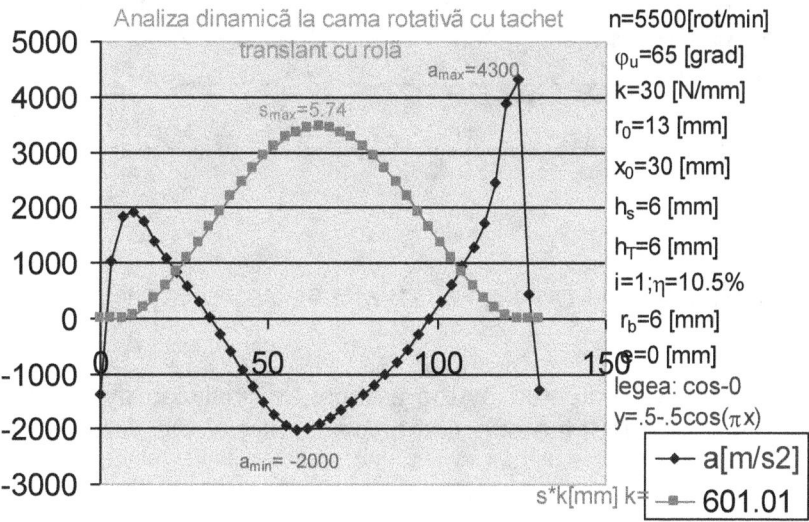

Fig. 4 *The dynamic analysis of the module B. Law COS, n=5500 rpm, φ_u=65^0, r_0=13 [mm], r_b=6 [mm], h_T=6 [mm], η=10.5%.*

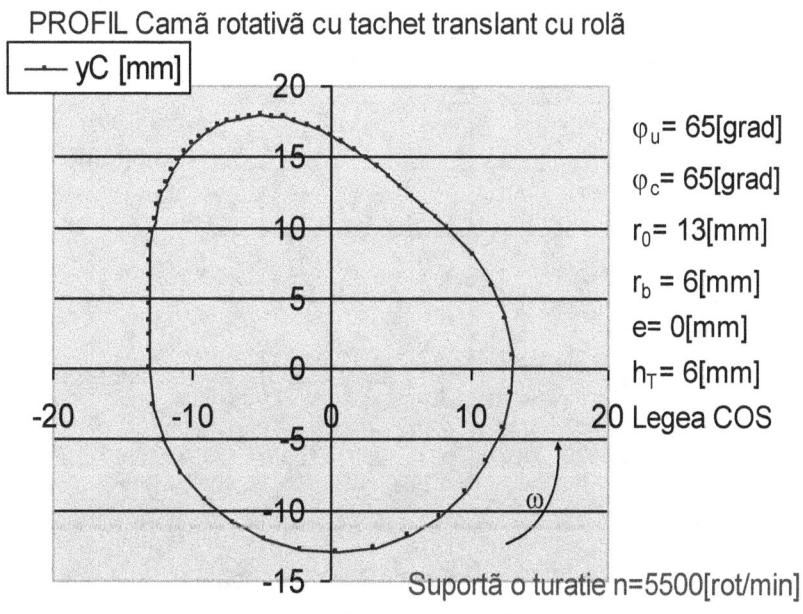

Fig. 5 *The profile COS at the module B, n=5500 rpm, φ_u=65^0, r_0=13 [mm], r_b=6 [mm], h_T=6 [mm].*

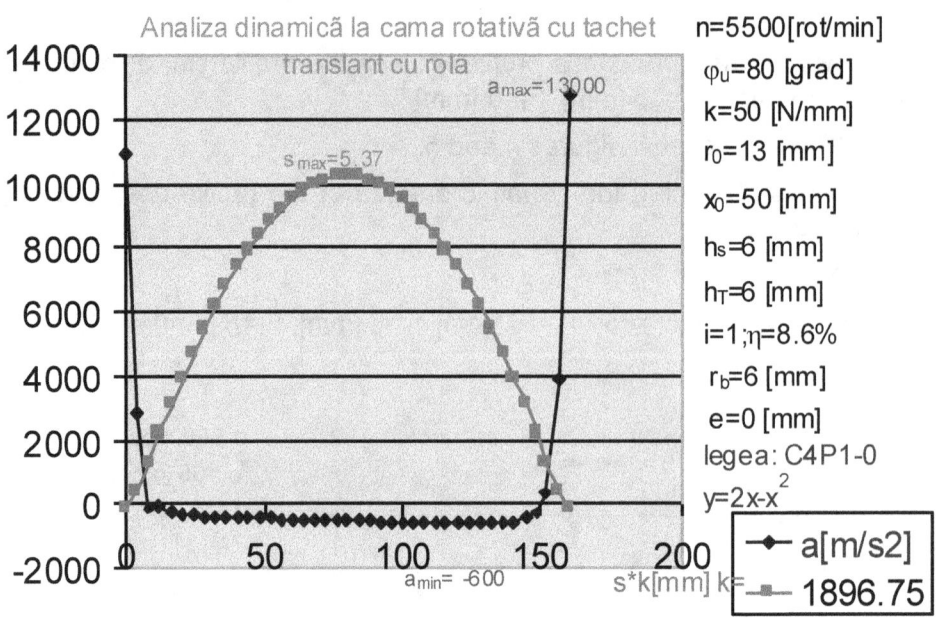

Fig. 6 *The dynamic analyze. Law C4P1-0, n=5500 rpm, $\varphi_u=80^0$, $r_0=13$ [mm], $r_b=6$ [mm], $h_T=6$ [mm].*

In figure 6 the law C4P, created by the authors, is analyzed dynamic. The vibrations are diminished, the noises are limited, the effective displacement of the valve is increased, $s_{max}=5.37$ [mm].

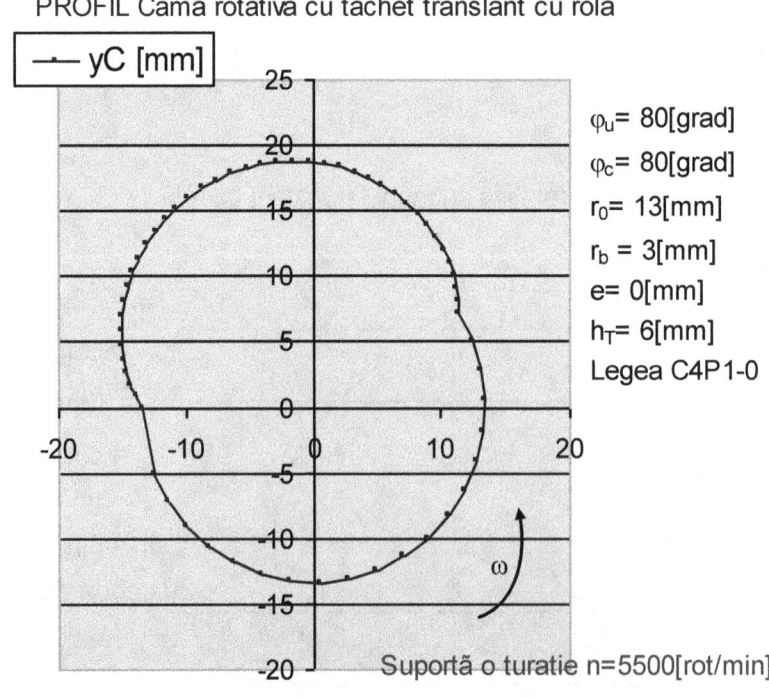

Fig. 7 *The profile C4P of the module B.*

The efficiency has a good value $\eta=8.6\%$. In figure 7 the profile of C4P law is presented. It starts at the law C4P with n=5500 [rpm], but for this law the rotation velocity can increase to high values of 30000-40000 [rpm] (see Fig. 8).

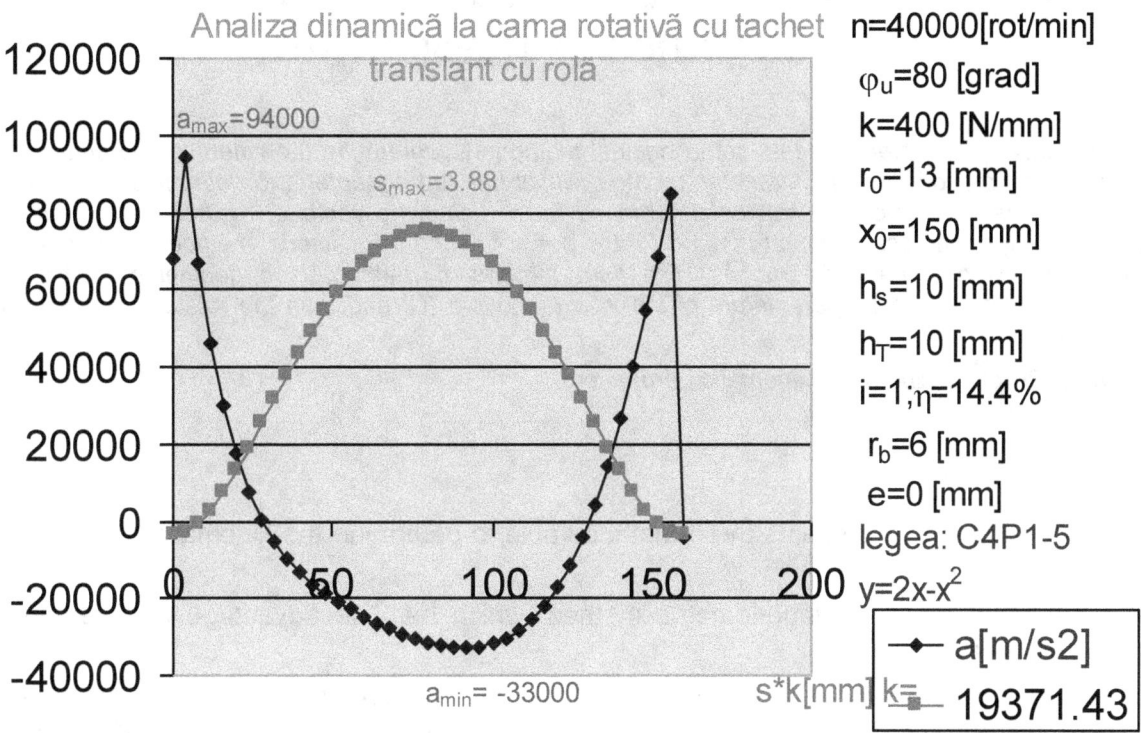

Fig. 8 *The dynamic analysis of the module B. Law C4P1-5, n=40000 rpm.*

3 Conclusions

We can speak about an advantage of the module B in comparison to the classical module C. With the module B, (when the follower is provided with a roll) it can obtain high rotation velocity with superior efficiency.

References

[1] Petrescu F.I., Petrescu R.V., *Contributions at the dynamics of cams*. In the Ninth IFToMM International Sympozium on Theory of Machines and Mechanisms, SYROM 2005, Bucharest, Romania, Vol. I, pp. 123-128, 2005.

CHAPTER IX
CINEMATICS OF THE 3R DYAD

Abstract: *This chapter presents some original methods to determine the kinematic parameters at the 3R dyad. It is starting with a trigonometric method, which has the advantage to determine very quickly the positions angles. The velocities can be determined faster using the vectorial method, so, to the second proposed method, one uses the first method for the positions, and the vectorial method for the determining of the velocities and the accelerations. The third (proposed) method, is a geometric method, which determine first the kinematic parameters of the internal couple (C) and then the rotation angles with their derivatives.*

Keywords: *3R dyad, cinematic, kinematic parameters*

1. Introduction

In this chapter it presents three methods able to determine the kinematic parameters to a 3R dyad (see the Figure 1).

It is starting with a trigonometric method, which has the advantage to determine very quickly the positions angles.

The velocities can be determined faster using the vectorial method, so, to the second proposed method, one uses the first method for the positions, and the vectorial method for the determining of the velocities and the accelerations.

The third (proposed) method, is a geometric method, which determine first the kinematic parameters of the internal couple (C) and then the rotation angles with their derivatives.

2. A trigonometric method

The kinematic schema of a RRR dyad can be seen in the Figure 1.

The following kinematic parameters considered known:

$x_B; y_B; x_D; y_D; \dot{x}_B; \dot{y}_B; \dot{x}_D; \dot{y}_D; \ddot{x}_B; \ddot{y}_B; \ddot{x}_D; \ddot{y}_D$

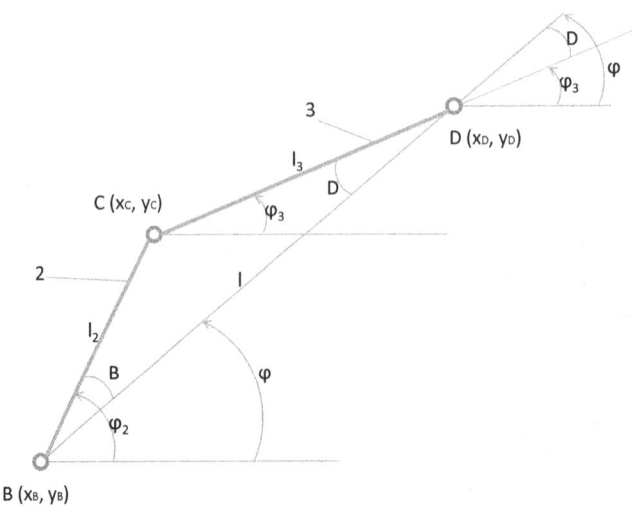

Fig. 1. *Kinematic schema of a 3R Dyad*

Must determine the parameters:

$\varphi_2, \dot{\varphi}_2, \ddot{\varphi}_2, \varphi_3, \dot{\varphi}_3, \ddot{\varphi}_3$

It determines the angle φ_2 and then the angle φ_3 in function of the three angles: $\varphi, \hat{B}, \hat{D}$, conform with the system (1).

$$\begin{cases} \varphi_2 = \varphi \pm \hat{B} \\ \varphi_3 = \varphi \mp \hat{D} \end{cases} \quad (1)$$

First, one calculates the length l between B and D (system 2).

$$\begin{cases} l^2 = (x_D - x_B)^2 + (y_D - y_B)^2 \\ l = \sqrt{(x_D - x_B)^2 + (y_D - y_B)^2} \end{cases} \quad (2)$$

The fi angle's parameters are determined with the system 3.

$$\begin{cases} \sin\varphi = \dfrac{y_D - y_B}{l}; \quad \cos\varphi = \dfrac{x_D - x_B}{l} \\ tg\varphi = \dfrac{y_D - y_B}{x_D - x_B}; \quad \varphi = sign(\sin\varphi) \cdot \arccos(\cos\varphi) \end{cases} \quad (3)$$

The velocity of the angle fi is writing with the relationship (4), and the acceleration is determined with the system (5).

$$\dot{\varphi} = \frac{(\dot{y}_D - \dot{y}_B) \cdot \cos\varphi - (\dot{x}_D - \dot{x}_B) \cdot \sin\varphi}{l} = \frac{(\dot{y}_D - \dot{y}_B) \cdot (x_D - x_B) - (\dot{x}_D - \dot{x}_B) \cdot (y_D - y_B)}{l^2} =$$

$$= \frac{(\dot{y}_D - \dot{y}_B) \cdot (x_D - x_B) - (\dot{x}_D - \dot{x}_B) \cdot (y_D - y_B)}{(x_D - x_B)^2 + (y_D - y_B)^2} \quad (4)$$

$$\begin{aligned} \dot{l} &= \frac{(x_D - x_B) \cdot (\dot{x}_D - \dot{x}_B) + (y_D - y_B) \cdot (\dot{y}_D - \dot{y}_B)}{l} \\ \ddot{\varphi} &= \frac{(\ddot{y}_D - \ddot{y}_B) \cdot \cos\varphi - (\ddot{x}_D - \ddot{x}_B) \cdot \sin\varphi - 2 \cdot \dot{l} \cdot \dot{\varphi}}{l} \end{aligned} \quad (5)$$

Next, we will determine kinematic parameters of the angle fi2 (system 6), and fi3 (system 7).

$$\begin{cases} \varphi_2 = \varphi \pm \hat{B}; \quad \cos B = \dfrac{l^2 + l_2^2 - l_3^2}{2 \cdot l \cdot l_2}; \quad \sin B = \dfrac{\sqrt{4 \cdot l^2 \cdot l_2^2 - (l^2 + l_2^2 - l_3^2)^2}}{2 \cdot l \cdot l_2} \\ \cos\varphi_2 = \cos(\varphi \pm B) = \cos\varphi \cdot \cos B \mp \sin\varphi \cdot \sin B \\ \sin\varphi_2 = \sin(\varphi \pm B) = \sin\varphi \cdot \cos B \pm \sin B \cdot \cos\varphi; \quad \varphi_2 = sign(\sin\varphi_2) \cdot \arccos(\cos\varphi_2) \\ 2 \cdot l_2 \cdot l \cdot \cos B = l_2^2 - l_3^2 + l^2; \quad l_2 \cdot l \cdot \sin B \cdot \dot{B} = l_2 \cdot \dot{l} \cdot \cos B - l \cdot \dot{l} \\ \dot{B} = \dfrac{l_2 \cdot \dot{l} \cdot \cos B - l \cdot \dot{l}}{l_2 \cdot l \cdot \sin B}; \quad \ddot{B} = \dfrac{l_2 \cdot \ddot{l} \cdot \cos B - 2 \cdot l_2 \cdot \dot{l} \cdot \sin B \cdot \dot{B} - l_2 \cdot l \cdot \cos B \cdot \dot{B}^2 - \dot{l}^2 - l \cdot \ddot{l}}{l_2 \cdot l \cdot \sin B} \\ \dot{\varphi}_2 = \dot{\varphi} \pm \dot{B}; \quad \ddot{\varphi}_2 = \ddot{\varphi} \pm \ddot{B} \end{cases} \quad (6)$$

$$\begin{cases} \varphi_3 = \varphi \mp D; \quad \cos D = \dfrac{l^2 + l_3^2 - l_2^2}{2 \cdot l \cdot l_3}; \quad \sin D = \dfrac{\sqrt{4 \cdot l^2 \cdot l_3^2 - (l^2 + l_3^2 - l_2^2)^2}}{2 \cdot l \cdot l_3} \\[6pt]
\cos \varphi_3 = \cos(\varphi \mp D) = \cos\varphi \cdot \cos D \pm \sin\varphi \cdot \sin D \\
\sin \varphi_3 = \sin(\varphi \mp D) = \sin\varphi \cdot \cos D \mp \sin D \cdot \cos\varphi; \quad \varphi_3 = sign(\sin\varphi_3) \cdot \arccos(\cos\varphi_3) \\[6pt]
2 \cdot l_3 \cdot l \cdot \cos D = l_3^2 - l_2^2 + l^2; \quad l_3 \cdot l \cdot \sin D \cdot \dot D = l_3 \cdot \dot l \cdot \cos D - l \cdot \dot l \\
\dot D = \dfrac{l_3 \cdot \dot l \cdot \cos D - l \cdot \dot l}{l_3 \cdot l \cdot \sin D}; \quad \ddot D = \dfrac{l_3 \cdot \ddot l \cdot \cos D - 2 \cdot l_3 \cdot \dot l \cdot \sin D \cdot \dot D - l_3 \cdot l \cdot \cos D \cdot \dot D^2 - \dot l^2 - l \cdot \ddot l}{l_3 \cdot l \cdot \sin D} \\[6pt]
\dot\varphi_3 = \dot\varphi \mp \dot D; \quad \ddot\varphi_3 = \ddot\varphi \mp \ddot D \end{cases} \quad (7)$$

Finally, one determines the parameters of point C (system 8).

$$\begin{cases} x_C = x_B + l_2 \cdot \cos\varphi_2 \\ y_C = y_B + l_2 \cdot \sin\varphi_2 \\ \dot x_C = \dot x_B - l_2 \cdot \sin\varphi_2 \cdot \omega_2 \\ \dot y_C = \dot y_B + l_2 \cdot \cos\varphi_2 \cdot \omega_2 \\ \ddot x_C = \ddot x_B - l_2 \cdot \cos\varphi_2 \cdot \omega_2^2 - l_2 \cdot \sin\varphi_2 \cdot \varepsilon_2 \\ \ddot y_C = \ddot y_B - l_2 \cdot \sin\varphi_2 \cdot \omega_2^2 + l_2 \cdot \cos\varphi_2 \cdot \varepsilon_2 \end{cases} \quad (8)$$

3. A combined method

The kinematic schema of a RRR dyad can be seen in the Figure 1.

The velocities can be determined faster using the vectorial method, so, to the second (proposed, combined) method, one uses the first method for the positions, and the vectorial method for the determining of the velocities and the accelerations. The most difficult problem at the 3R dyad is the determining of the positions. To eliminate the traditional processes, to which we need two times to rises squared the system equations, we will determine the positions with the direct relationships (9).

$$\begin{cases} \sin\varphi = \dfrac{y_D - y_B}{l}; \quad \cos\varphi = \dfrac{x_D - x_B}{l}; \\ \varphi = sign(\sin\varphi) \cdot \arccos(\cos\varphi) \\[6pt]
\cos B = \dfrac{l^2 + l_2^2 - l_3^2}{2 \cdot l \cdot l_2}; \quad B = \arccos(\cos B) \\[6pt]
\cos D = \dfrac{l^2 + l_3^2 - l_2^2}{2 \cdot l \cdot l_3}; \quad D = \arccos(\cos D) \end{cases} \quad (9)$$

For the determination of velocities and accelerations one uses the classical vectorial method (systems 10-11).

$$\begin{cases} l_2 \cdot \cos\varphi_2 + l_3 \cdot \cos\varphi_3 = x_D - x_B \\ l_2 \cdot \sin\varphi_2 + l_3 \cdot \sin\varphi_3 = y_D - y_B \end{cases} \Rightarrow \begin{cases} -l_2 \cdot \sin\varphi_2 \cdot \omega_2 - l_3 \cdot \sin\varphi_3 \cdot \omega_3 = \dot{x}_D - \dot{x}_B \\ l_2 \cdot \cos\varphi_2 \cdot \omega_2 + l_3 \cdot \cos\varphi_3 \cdot \omega_3 = \dot{y}_D - \dot{y}_B \end{cases}$$

$$\begin{cases} -l_2 \cdot \sin\varphi_2 \cdot \omega_2 - l_3 \cdot \sin\varphi_3 \cdot \omega_3 = \dot{x}_D - \dot{x}_B \,|\cdot(\cos\varphi_3) \\ l_2 \cdot \cos\varphi_2 \cdot \omega_2 + l_3 \cdot \cos\varphi_3 \cdot \omega_3 = \dot{y}_D - \dot{y}_B \,|\cdot(\sin\varphi_3) \end{cases} |+ \Rightarrow \omega_2 = \frac{(\dot{x}_D - \dot{x}_B)\cdot\cos\varphi_3 + (\dot{y}_D - \dot{y}_B)\cdot\sin\varphi_3}{l_2 \cdot \sin(\varphi_3 - \varphi_2)} \quad (10)$$

$$\begin{cases} -l_2 \cdot \sin\varphi_2 \cdot \omega_2 - l_3 \cdot \sin\varphi_3 \cdot \omega_3 = \dot{x}_D - \dot{x}_B \,|\cdot(\cos\varphi_2) \\ l_2 \cdot \cos\varphi_2 \cdot \omega_2 + l_3 \cdot \cos\varphi_3 \cdot \omega_3 = \dot{y}_D - \dot{y}_B \,|\cdot(\sin\varphi_2) \end{cases} |+ \Rightarrow \omega_3 = \frac{(\dot{x}_D - \dot{x}_B)\cdot\cos\varphi_2 + (\dot{y}_D - \dot{y}_B)\cdot\sin\varphi_2}{l_3 \cdot \sin(\varphi_2 - \varphi_3)}$$

$$\begin{cases} -l_2 \cdot \sin\varphi_2 \cdot \omega_2 - l_3 \cdot \sin\varphi_3 \cdot \omega_3 = \dot{x}_D - \dot{x}_B \\ l_2 \cdot \cos\varphi_2 \cdot \omega_2 + l_3 \cdot \cos\varphi_3 \cdot \omega_3 = \dot{y}_D - \dot{y}_B \end{cases} \Rightarrow \begin{cases} -l_2 \cdot \cos\varphi_2 \cdot \omega_2^2 - l_2 \cdot \sin\varphi_2 \cdot \varepsilon_2 - \\ -l_3 \cdot \cos\varphi_3 \cdot \omega_3^2 - l_3 \cdot \sin\varphi_3 \cdot \varepsilon_3 = \ddot{x}_D - \ddot{x}_B \\ -l_2 \cdot \sin\varphi_2 \cdot \omega_2^2 + l_2 \cdot \cos\varphi_2 \cdot \varepsilon_2 - \\ -l_3 \cdot \sin\varphi_3 \cdot \omega_3^2 + l_3 \cdot \cos\varphi_3 \cdot \varepsilon_3 = \ddot{y}_D - \ddot{y}_B \end{cases}$$

$$\begin{cases} -l_2 \cdot \cos\varphi_2 \cdot \omega_2^2 - l_2 \cdot \sin\varphi_2 \cdot \varepsilon_2 - l_3 \cdot \cos\varphi_3 \cdot \omega_3^2 - \\ -l_3 \cdot \sin\varphi_3 \cdot \varepsilon_3 = \ddot{x}_D - \ddot{x}_B \,|\cdot(\cos\varphi_3) \\ -l_2 \cdot \sin\varphi_2 \cdot \omega_2^2 + l_2 \cdot \cos\varphi_2 \cdot \varepsilon_2 - l_3 \cdot \sin\varphi_3 \cdot \omega_3^2 + \\ + l_3 \cdot \cos\varphi_3 \cdot \varepsilon_3 = \ddot{y}_D - \ddot{y}_B \,|\cdot(\sin\varphi_3) \end{cases} |+ \Rightarrow \varepsilon_2 = \frac{(\ddot{x}_D - \ddot{x}_B)\cos\varphi_3 + (\ddot{y}_D - \ddot{y}_B)\sin\varphi_3 + l_2\omega_2^2\cos(\varphi_3 - \varphi_2) + l_3\omega_3^2}{l_2 \cdot \sin(\varphi_3 - \varphi_2)}$$

$$\begin{cases} -l_2 \cdot \cos\varphi_2 \cdot \omega_2^2 - l_2 \cdot \sin\varphi_2 \cdot \varepsilon_2 - l_3 \cdot \cos\varphi_3 \cdot \omega_3^2 - \\ -l_3 \cdot \sin\varphi_3 \cdot \varepsilon_3 = \ddot{x}_D - \ddot{x}_B \,|\cdot(\cos\varphi_2) \\ -l_2 \cdot \sin\varphi_2 \cdot \omega_2^2 + l_2 \cdot \cos\varphi_2 \cdot \varepsilon_2 - l_3 \cdot \sin\varphi_3 \cdot \omega_3^2 + \\ + l_3 \cdot \cos\varphi_3 \cdot \varepsilon_3 = \ddot{y}_D - \ddot{y}_B \,|\cdot(\sin\varphi_2) \end{cases} |+ \Rightarrow \varepsilon_3 = \frac{(\ddot{x}_D - \ddot{x}_B)\cos\varphi_2 + (\ddot{y}_D - \ddot{y}_B)\sin\varphi_2 + l_2\omega_2^2 + l_3\omega_3^2\cos(\varphi_2 - \varphi_3)}{l_3 \cdot \sin(\varphi_2 - \varphi_3)} \quad (11)$$

4. A geometric method

The kinematic schema of a RRR dyad can be seen in the Figure 1.

The third (proposed) method, is a geometric method, which determine first the kinematic parameters of the internal couple (C) and then the rotation angles with their derivatives.

We start with the geometric positions (the system 12).

$$\begin{cases} (x - x_B)^2 + (y - y_B)^2 = l_2^2 \\ (x - x_D)^2 + (y - y_D)^2 = l_3^2 \end{cases} \quad (12)$$

These equations were deduced geometrically, by the writing of two equations for a two circles (x=x_C, y=y_C).

To solving the system (12) it writing the system (13).

$$\begin{cases}
(y-y_B)^2 = l_2^2 - (x-x_B)^2 \\
x - x_D = \pm\sqrt{l_3^2 - (y-y_D)^2}; \; x = x_D \pm \sqrt{l_3^2 - (y-y_D)^2}; \; x - x_B = (x_D - x_B) \pm \sqrt{l_3^2 - (y-y_D)^2} \\
(x-x_B)^2 = (x_D - x_B)^2 + \left[l_3^2 - (y-y_D)^2\right] \pm 2 \cdot (x_D - x_B) \cdot \sqrt{l_3^2 - (y-y_D)^2} \\
(x-x_B)^2 = (x_D - x_B)^2 + l_3^2 - (y-y_D)^2 \pm 2 \cdot (x_D - x_B) \cdot \sqrt{l_3^2 - (y-y_D)^2} \\
\\
(y-y_B)^2 = l_2^2 - (x_D - x_B)^2 - l_3^2 + (y-y_D)^2 \mp 2 \cdot (x_D - x_B) \cdot \sqrt{l_3^2 - (y-y_D)^2} \\
y^2 + y_B^2 - 2 \cdot y_B \cdot y = l_2^2 - (x_D - x_B)^2 - l_3^2 + y^2 + y_D^2 - 2 \cdot y_D \cdot y \mp \\
\mp 2 \cdot (x_D - x_B) \cdot \sqrt{l_3^2 - (y-y_D)^2} \\
2 \cdot (y_D - y_B) \cdot y + \left[y_B^2 - l_2^2 + (x_D - x_B)^2 + l_3^2 - y_D^2\right] = \mp 2 \cdot (x_D - x_B) \cdot \sqrt{l_3^2 - (y-y_D)^2} \\
2 \cdot (y_D - y_B) = b; \; y_B^2 - l_2^2 + (x_D - x_B)^2 + l_3^2 - y_D^2 = d; \; 2 \cdot (x_D - x_B) = a \\
b \cdot y + d = \mp a \cdot \sqrt{l_3^2 - (y-y_D)^2} \\
b^2 \cdot y^2 + d^2 + 2 \cdot b \cdot d \cdot y = a^2 \cdot l_3^2 - a^2 \cdot y^2 - a^2 \cdot y_D^2 + 2 \cdot a^2 \cdot y_D \cdot y \\
(a^2 + b^2) \cdot y^2 - 2 \cdot (a^2 \cdot y_D - b \cdot d) \cdot y - (a^2 \cdot l_3^2 - a^2 \cdot y_D^2 - d^2) = 0 \\
\Delta(R) = (a^2 \cdot y_D - b \cdot d)^2 + (a^2 + b^2) \cdot (a^2 \cdot l_3^2 - a^2 \cdot y_D^2 - d^2) = \\
= a^4 \cdot y_D^2 - a^4 \cdot y_D^2 + b^2 \cdot d^2 - b^2 \cdot d^2 - 2 \cdot a^2 \cdot b \cdot d \cdot y_D - \\
- a^2 \cdot d^2 + a^4 \cdot l_3^2 + a^2 \cdot b^2 \cdot l_3^2 - a^2 \cdot b^2 \cdot y_D^2 = \\
= a^2 \cdot \left[l_3^2 \cdot (a^2 + b^2) - (d + b \cdot y_D)^2\right] \\
\\
c = x_B^2 - x_D^2 + y_B^2 - y_D^2 + l_3^2 - l_2^2 \\
y_{1,2} = \dfrac{a^2 \cdot y_D - b \cdot d \pm a \cdot \sqrt{l_3^2 \cdot (a^2 + b^2) - (d + b \cdot y_D)^2}}{a^2 + b^2}; \; x_{1,2} = -\dfrac{b}{a} \cdot y_{1,2} - \dfrac{c}{a} \\
\\
+ \text{when } C \text{ at North} \quad - \text{when } C \text{ at South} \\
\begin{cases} y_C \equiv y = \dfrac{a^2 \cdot y_D - b \cdot d + a \cdot \sqrt{l_3^2 \cdot (a^2 + b^2) - (d + b \cdot y_D)^2}}{a^2 + b^2} \\ x_C \equiv x = -\dfrac{b}{a} \cdot y - \dfrac{c}{a} \end{cases}
\end{cases} \quad (13)$$

$$\begin{cases} (x-x_B)^2 + (y-y_B)^2 = l_2^2 \\ (x-x_D)^2 + (y-y_D)^2 = l_3^2 \end{cases} \quad (12)$$

To determine the velocities and the accelerations we derivate the equations system (12) and we have obtained the system (14).

$$\begin{cases}
2\cdot(x-x_B)\cdot(\dot{x}-\dot{x}_B)+2\cdot(y-y_B)\cdot(\dot{y}-\dot{y}_B)=0 \\
2\cdot(x-x_D)\cdot(\dot{x}-\dot{x}_D)+2\cdot(y-y_D)\cdot(\dot{y}-\dot{y}_D)=0 \\
(x-x_B)\cdot\dot{x}+(y-y_B)\cdot\dot{y}=(x-x_B)\cdot\dot{x}_B+(y-y_B)\cdot\dot{y}_B \\
(x-x_D)\cdot\dot{x}+(y-y_D)\cdot\dot{y}=(x-x_D)\cdot\dot{x}_D+(y-y_D)\cdot\dot{y}_D \\
a_{11}=x-x_B;\ a_{12}=y-y_B;\ b_1=(x-x_B)\cdot\dot{x}_B+(y-y_B)\cdot\dot{y}_B \\
a_{21}=x-x_D;\ a_{22}=y-y_D;\ b_2=(x-x_D)\cdot\dot{x}_D+(y-y_D)\cdot\dot{y}_D \\
\Delta=\begin{vmatrix}a_{11} & a_{12}\\ a_{21} & a_{22}\end{vmatrix}=a_{11}\cdot a_{22}-a_{21}\cdot a_{12};\ \Delta_{\dot{x}}=\begin{vmatrix}b_1 & a_{12}\\ b_2 & a_{22}\end{vmatrix}=b_1\cdot a_{22}-b_2\cdot a_{12}; \\
\Delta_{\dot{y}}=\begin{vmatrix}a_{11} & b_1\\ a_{21} & b_2\end{vmatrix}=a_{11}\cdot b_2-a_{21}\cdot b_1;\ \dot{x}\equiv\dot{x}_C=\frac{\Delta_{\dot{x}}}{\Delta};\ \dot{y}\equiv\dot{y}_C=\frac{\Delta_{\dot{y}}}{\Delta} \\[6pt]
(\dot{x}-\dot{x}_B)\cdot\dot{x}+(x-x_B)\cdot\ddot{x}+(\dot{y}-\dot{y}_B)\cdot\dot{y}+(y-y_B)\cdot\ddot{y}= \\
=(\dot{x}-\dot{x}_B)\cdot\dot{x}_B+(x-x_B)\cdot\ddot{x}_B+(\dot{y}-\dot{y}_B)\cdot\dot{y}_B+(y-y_B)\cdot\ddot{y}_B \\
(\dot{x}-\dot{x}_D)\cdot\dot{x}+(x-x_D)\cdot\ddot{x}+(\dot{y}-\dot{y}_D)\cdot\dot{y}+(y-y_D)\cdot\ddot{y}= \\
=(\dot{x}-\dot{x}_D)\cdot\dot{x}_D+(x-x_D)\cdot\ddot{x}_D+(\dot{y}-\dot{y}_D)\cdot\dot{y}_D+(y-y_D)\cdot\ddot{y}_D \\
\begin{cases}a_{11}\cdot\ddot{x}+a_{12}\cdot\ddot{y}=c_1\\ a_{21}\cdot\ddot{x}+a_{22}\cdot\ddot{y}=c_2\end{cases} \\
\begin{cases}c_1=(x-x_B)\cdot\ddot{x}_B+(y-y_B)\cdot\ddot{y}_B-(\dot{x}-\dot{x}_B)^2-(\dot{y}-\dot{y}_B)^2\\ c_2=(x-x_D)\cdot\ddot{x}_D+(y-y_D)\cdot\ddot{y}_D-(\dot{x}-\dot{x}_D)^2-(\dot{y}-\dot{y}_D)^2\end{cases} \\
\Delta_{\ddot{x}}=\begin{vmatrix}c_1 & a_{12}\\ c_2 & a_{22}\end{vmatrix}=a_{22}\cdot c_1-a_{12}\cdot c_2;\ \ddot{x}\equiv\ddot{x}_C=\frac{\Delta_{\ddot{x}}}{\Delta}; \\
\Delta_{\ddot{y}}=\begin{vmatrix}a_{11} & c_1\\ a_{21} & c_2\end{vmatrix}=a_{11}\cdot c_2-a_{21}\cdot c_1;\ \ddot{y}\equiv\ddot{y}_C=\frac{\Delta_{\ddot{y}}}{\Delta}
\end{cases} \qquad (14)$$

Finally one determines the positions angles and their derivatives (15).

$$\begin{cases}
\begin{cases} x_C = x_B + l_2 \cdot \cos\varphi_2 \\ y_C = y_B + l_2 \cdot \sin\varphi_2 \end{cases} \quad \begin{cases} x_D = x_C + l_3 \cdot \cos\varphi_3 \\ y_D = y_C + l_3 \cdot \sin\varphi_3 \end{cases} \\
\begin{cases} x_C - x_B = l_2 \cdot \cos\varphi_2 \\ y_C - y_B = l_2 \cdot \sin\varphi_2 \end{cases} \quad \begin{cases} x_D - x_C = l_3 \cdot \cos\varphi_3 \\ y_D - y_C = l_3 \cdot \sin\varphi_3 \end{cases} \\
\cos\varphi_2 = \dfrac{x_C - x_B}{l_2}; \quad \sin\varphi_2 = \dfrac{y_C - y_B}{l_2}; \\
\cos\varphi_3 = \dfrac{x_D - x_C}{l_3}; \quad \sin\varphi_3 = \dfrac{y_D - y_C}{l_3} \\
\begin{cases} \dot{x}_C - \dot{x}_B = -l_2 \cdot \sin\varphi_2 \cdot \omega_2 \mid \cdot(-\sin\varphi_2) \\ \dot{y}_C - \dot{y}_B = l_2 \cdot \cos\varphi_2 \cdot \omega_2 \mid \cdot(\cos\varphi_2) \end{cases} \Rightarrow \\
\Rightarrow \omega_2 = \dfrac{(\dot{y}_C - \dot{y}_B) \cdot \cos\varphi_2 - (\dot{x}_C - \dot{x}_B) \cdot \sin\varphi_2}{l_2} \\
\begin{cases} \dot{x}_D - \dot{x}_C = -l_3 \cdot \sin\varphi_3 \cdot \omega_3 \mid \cdot(-\sin\varphi_3) \\ \dot{y}_D - \dot{y}_C = l_3 \cdot \cos\varphi_3 \cdot \omega_3 \mid \cdot(\cos\varphi_3) \end{cases} \Rightarrow \\
\Rightarrow \omega_3 = \dfrac{(\dot{y}_D - \dot{y}_C) \cdot \cos\varphi_3 - (\dot{x}_D - \dot{x}_C) \cdot \sin\varphi_3}{l_3} \\
\begin{cases} \ddot{x}_C - \ddot{x}_B = -l_2 \cos\varphi_2 \cdot \omega_2^2 - l_2 \sin\varphi_2 \cdot \varepsilon_2 \mid -\sin\varphi_2 \\ \ddot{y}_C - \ddot{y}_B = -l_2 \sin\varphi_2 \cdot \omega_2^2 + l_2 \cos\varphi_2 \cdot \varepsilon_2 \mid \cos\varphi_2 \end{cases} \Rightarrow \\
\Rightarrow \varepsilon_2 = \dfrac{(\ddot{y}_C - \ddot{y}_B) \cdot \cos\varphi_2 - (\ddot{x}_C - \ddot{x}_B) \cdot \sin\varphi_2}{l_2} \\
\begin{cases} \ddot{x}_D - \ddot{x}_C = -l_3 \cos\varphi_3 \cdot \omega_3^2 - l_3 \sin\varphi_3 \cdot \varepsilon_3 \mid -\sin\varphi_3 \\ \ddot{y}_D - \ddot{y}_C = -l_3 \sin\varphi_3 \cdot \omega_3^2 + l_3 \cos\varphi_3 \cdot \varepsilon_3 \mid \cos\varphi_3 \end{cases} \Rightarrow \\
\Rightarrow \varepsilon_3 = \dfrac{(\ddot{y}_D - \ddot{y}_C) \cdot \cos\varphi_3 - (\ddot{x}_D - \ddot{x}_C) \cdot \sin\varphi_3}{l_3}
\end{cases} \quad (15)$$

Conclusions

The geometrical method presented in the last paragraph is the most elegant and direct method to determine the positions angles and their derivatives.

Relationships used by this method allow and the determination of the system vibrations.

References

[1] Pelecudi, Chr., ş.a., *Mecanisme*. E.D.P., Bucureşti, 1985.

CHAPTER X

KINEMATICS OF THE PLANAR QUADRILATERAL MECHANISM

Abstract: This chapter presents an original method to determine the kinematic parameters at the linked quadrilateral mechanism. It is starting with a trigonometric method, which has the advantage to determine very quickly the position angles. The velocities can be determined faster using a geometric method. This method is developed and for the accelerations determinations. The (proposed) geometric method, determines first the kinematic parameters of the internal couple (B) and then the rotation angles with their derivatives. Secondary, the chapter presents the determination of the efficiency of this mechanism. Determines and dynamic coefficient, D. With this one proposes two yields; the mechanical efficiency and the dynamic efficiency.
Keywords: 3R dyad, cinematic, kinematic parameters, efficiency, dynamic coefficient

1. Introduction

The chapter presents an original method to determine the kinematic parameters at the linked quadrilateral mechanism.

It is starting with a trigonometric method, which has the advantage to determine very quickly the position angles.

The velocities can be determined faster using a geometric method. This method is developed and for the accelerations determinations. The (proposed) geometric method, determines first the kinematic parameters of the internal couple (B) and then the rotation angles with their derivatives.

Secondary, the chapter presents the determination of the efficiency of this mechanism. Determines and dynamic coefficient, D.

With this one proposes two yields; the mechanical efficiency and the dynamic efficiency.

2. The kinematics of the planar quadrilateral mechanism

The kinematic schema of a planar quadrilateral mechanism can be seen in the Figure 1.

The following kinematic parameters considered known:

$x_O; y_O; x_C; y_C; l_1; l_2; l_3; \varphi_1; \omega_1 = ct.$

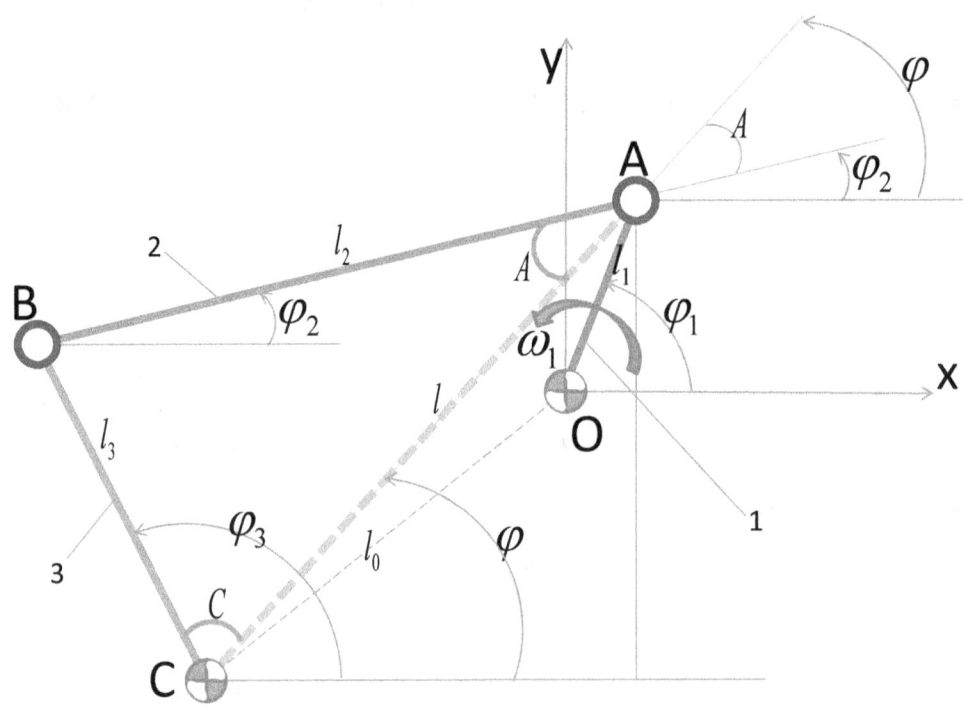

Fig. 1. *Kinematic schema of a planar quadrilateral mechanism*

2.1. Determining the positions

It is starting with a trigonometric method, which has the advantage to determine very quickly the position angles (the system 1).

$$\begin{cases} \begin{cases} x_A = l_1 \cdot \cos \varphi_1 \\ y_A = l_1 \cdot \sin \varphi_1 \end{cases} \begin{cases} \dot{x}_A = -l_1 \cdot \sin \varphi_1 \cdot \omega_1 \\ \dot{y}_A = l_1 \cdot \cos \varphi_1 \cdot \omega_1 \end{cases} \begin{cases} \ddot{x}_A = -l_1 \cdot \cos \varphi_1 \cdot \omega_1^2 \\ \ddot{y}_A = -l_1 \cdot \sin \varphi_1 \cdot \omega_1^2 \end{cases} \\ l^2 = (x_A - x_C)^2 + (y_A - y_C)^2 \Rightarrow l = \sqrt{l^2} = \sqrt{(x_A - x_C)^2 + (y_A - y_C)^2} \\ \begin{cases} \cos A = \dfrac{l^2 + l_2^2 - l_3^2}{2 \cdot l \cdot l_2} \Rightarrow A = \arccos(\cos A); \\ \cos C = \dfrac{l^2 + l_3^2 - l_2^2}{2 \cdot l \cdot l_3} \Rightarrow C = \arccos(\cos C) \end{cases} \begin{cases} \cos \varphi = \dfrac{x_A - x_C}{l} \\ \sin \varphi = \dfrac{y_A - y_C}{l} \end{cases} \Rightarrow \\ \varphi = sign(\sin \varphi) \cdot \arccos(\cos \varphi) \\ \begin{cases} \varphi_2 = \varphi - A \\ \varphi_3 = \varphi + C \end{cases} \begin{cases} x_B = x_C + l_3 \cdot \cos \varphi_3 \\ y_B = y_C + l_3 \cdot \sin \varphi_3 \end{cases} \end{cases} \quad (1)$$

2.2. Determining the velocities of the couple B

The velocities can be determined faster using a geometric method (the system 2).

$$\begin{cases} \begin{cases} (x_B - x_C)^2 + (y_B - y_C)^2 = l_3^2 \\ (x_B - x_A)^2 + (y_B - y_A)^2 = l_2^2 \end{cases} \\[1em] \begin{cases} (x_B - x_C) \cdot \dot{x}_B + (y_B - y_C) \cdot \dot{y}_B = 0 \\ (x_B - x_A) \cdot \dot{x}_B + (y_B - y_A) \cdot \dot{y}_B = (x_B - x_A) \cdot \dot{x}_A + (y_B - y_A) \cdot \dot{y}_A \end{cases} \\[1em] a_{11} = x_B - x_C; \quad a_{12} = y_B - y_C; \quad a_{21} = x_B - x_A; \quad a_{22} = y_B - y_A; \\[0.5em] b_1 = 0; \quad b_2 = a_{21} \cdot \dot{x}_A + a_{22} \cdot \dot{y}_A \\[1em] \begin{cases} a_{11} \cdot \dot{x}_B + a_{12} \cdot \dot{y}_B = b_1 \\ a_{21} \cdot \dot{x}_B + a_{22} \cdot \dot{y}_B = b_2 \end{cases} \\[1em] \Delta = \begin{vmatrix} a_{11} & a_{12} \\ a_{21} & a_{22} \end{vmatrix} = a_{11} \cdot a_{22} - a_{12} \cdot a_{21}; \\[1em] \Delta_{\dot{x}_B} = \begin{vmatrix} b_1 & a_{12} \\ b_2 & a_{22} \end{vmatrix} = b_1 \cdot a_{22} - a_{12} \cdot b_2 \\[1em] \Delta_{\dot{y}_B} = \begin{vmatrix} a_{11} & b_1 \\ a_{21} & b_2 \end{vmatrix} = a_{11} \cdot b_2 - a_{21} \cdot b_1; \\[1em] \dot{x}_B = \frac{\Delta_{\dot{x}_B}}{\Delta}; \quad \dot{y}_B = \frac{\Delta_{\dot{y}_B}}{\Delta} \end{cases} \quad (2)$$

2.3. Determining the accelerations of the couple B

The accelerations can be determined faster using a geometric method (the system 3).

$$\begin{cases} \begin{cases} (x_B - x_C)\cdot \ddot{x}_B + (y_B - y_C)\cdot \ddot{y}_B = -\dot{x}_B^2 - \dot{y}_B^2 \\ (x_B - x_A)\cdot \ddot{x}_B + (y_B - y_A)\cdot \ddot{y}_B = a_{21}\cdot \ddot{x}_A + a_{22}\cdot \ddot{y}_A - \dot{a}_{21}^2 - \dot{a}_{22}^2 \end{cases} \\ c_1 = -\dot{x}_B^2 - \dot{y}_B^2; \quad c_2 = a_{21}\cdot \ddot{x}_A + a_{22}\cdot \ddot{y}_A - \dot{a}_{21}^2 - \dot{a}_{22}^2 \\ \begin{cases} a_{11}\cdot \ddot{x}_B + a_{12}\cdot \ddot{y}_B = c_1 \\ a_{21}\cdot \ddot{x}_B + a_{22}\cdot \ddot{y}_B = c_2 \end{cases} \Rightarrow \Delta_{\ddot{x}_B} = \begin{vmatrix} c_1 & a_{12} \\ c_2 & a_{22} \end{vmatrix} = c_1\cdot a_{22} - a_{12}\cdot c_2 \\ \Delta_{\ddot{y}_B} = \begin{vmatrix} a_{11} & c_1 \\ a_{21} & c_2 \end{vmatrix} = a_{11}\cdot c_2 - a_{21}\cdot c_1; \quad \ddot{x}_B = \frac{\Delta_{\ddot{x}_B}}{\Delta}; \quad \ddot{y}_B = \frac{\Delta_{\ddot{y}_B}}{\Delta} \end{cases} \quad (3)$$

2.4. Determining the angular velocities and accelerations

The angular velocities and accelerations can be determined now, faster, using the vectorial method (the system 4).

$$\begin{cases} \begin{cases} x_A - x_B = l_2\cdot \cos\varphi_2 \\ y_A - y_B = l_2\cdot \sin\varphi_2 \end{cases} \begin{cases} \dot{x}_A - \dot{x}_B = -l_2\cdot \sin\varphi_2\cdot \omega_2\,|\cdot(-\sin\varphi_2) \\ \dot{y}_A - \dot{y}_B = l_2\cdot \cos\varphi_2\cdot \omega_2\,|\cdot(\cos\varphi_2) \end{cases} \Rightarrow \omega_2 = \frac{(\dot{y}_A - \dot{y}_B)\cdot \cos\varphi_2 - (\dot{x}_A - \dot{x}_B)\cdot \sin\varphi_2}{l_2} \\ \\ \begin{cases} \ddot{x}_A - \ddot{x}_B = -l_2\cdot \sin\varphi_2\cdot \varepsilon_2 - l_2\cdot \cos\varphi_2\cdot \omega_2^2\,|\cdot(-\sin\varphi_2) \\ \ddot{y}_A - \ddot{y}_B = l_2\cdot \cos\varphi_2\cdot \varepsilon_2 - l_2\cdot \sin\varphi_2\cdot \omega_2^2\,|\cdot(\cos\varphi_2) \end{cases} \Rightarrow \varepsilon_2 = \frac{(\ddot{y}_A - \ddot{y}_B)\cdot \cos\varphi_2 - (\ddot{x}_A - \ddot{x}_B)\cdot \sin\varphi_2}{l_2} \\ \\ \begin{cases} x_B - x_C = l_3\cdot \cos\varphi_3 \\ y_B - y_C = l_3\cdot \sin\varphi_3 \end{cases} \begin{cases} \dot{x}_B - \dot{x}_C = -l_3\cdot \sin\varphi_3\cdot \omega_3\,|\cdot(-\sin\varphi_3) \\ \dot{y}_B - \dot{y}_C = l_3\cdot \cos\varphi_3\cdot \omega_3\,|\cdot(\cos\varphi_3) \end{cases} \Rightarrow \omega_3 = \frac{(\dot{y}_B - \dot{y}_C)\cdot \cos\varphi_3 - (\dot{x}_B - \dot{x}_C)\cdot \sin\varphi_3}{l_3} \\ \\ \begin{cases} \ddot{x}_B - \ddot{x}_C = -l_3\cdot \sin\varphi_3\cdot \varepsilon_3 - l_3\cdot \cos\varphi_3\cdot \omega_3^2\,|\cdot(-\sin\varphi_3) \\ \ddot{y}_B - \ddot{y}_C = l_3\cdot \cos\varphi_3\cdot \varepsilon_3 - l_3\cdot \sin\varphi_3\cdot \omega_3^2\,|\cdot(\cos\varphi_3) \end{cases} \Rightarrow \varepsilon_3 = \frac{(\ddot{y}_B - \ddot{y}_C)\cdot \cos\varphi_3 - (\ddot{x}_B - \ddot{x}_C)\cdot \sin\varphi_3}{l_3} \end{cases} \quad (4)$$

3. The efficiency of the planar quadrilateral mechanism

The efficiency of a planar quadrilateral mechanism can be determined starting from the forces and velocities repartition, (Figure 2).

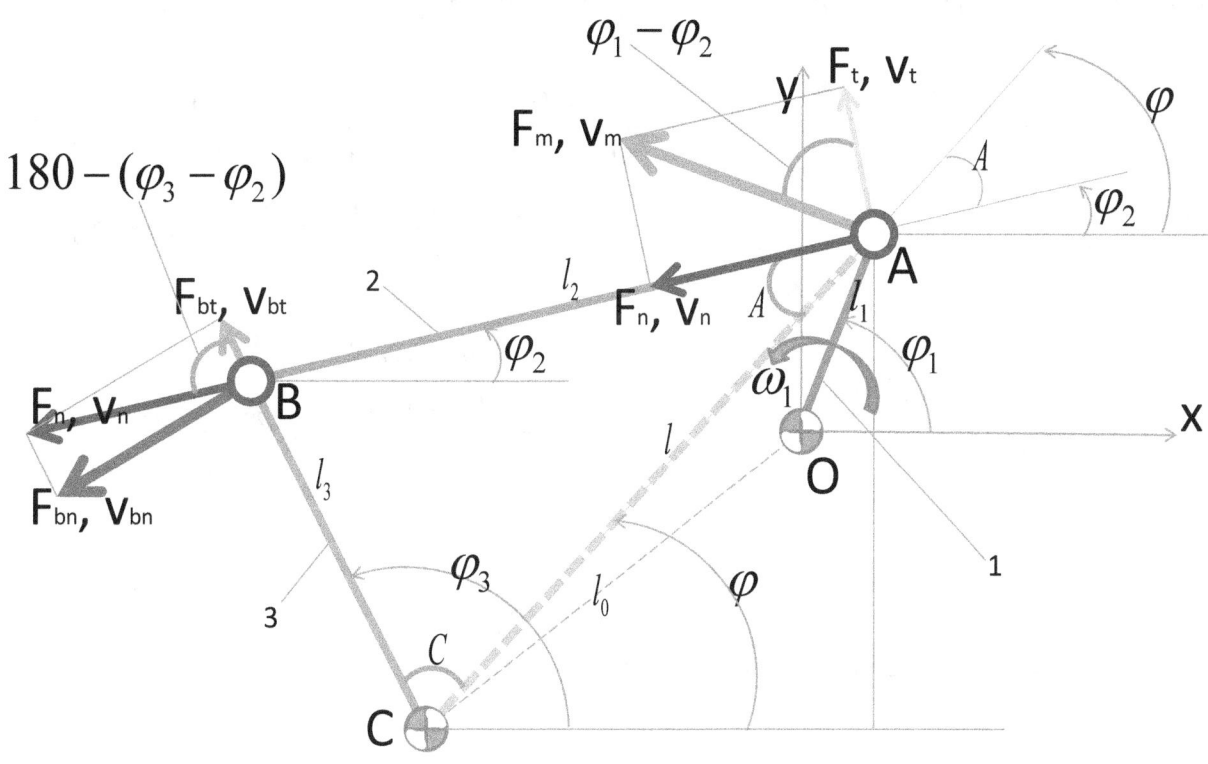

Fig. 2. *Forces and velocities repartition of a planar quadrilateral mechanism*

The system (5) presents the relationships which give the forces and the velocities on the planar quadrilateral mechanism. The driving force F_m is perpendicular on the crank 1 in A.

Its component along the connecting rod (the bar 2) F_n, gives the normal component F_{bn}. F_{bn} is perpendicular on the rocker 3 in B.

These forces give the dynamic velocities which are similar with the forces.

The forces are always the same, but the velocities (the dynamic velocities) are different than the kinematics velocities.

For this reason the dynamic efficiency will be different than the mechanical yield.

77

$$\begin{cases} \begin{cases} F_n = F_m \cdot \sin(\varphi_1 - \varphi_2) \\ v_n = v_m \cdot \sin(\varphi_1 - \varphi_2) \end{cases} \\ \begin{cases} F_B \equiv F_{bn} = F_n \cdot \sin[\pi - (\varphi_3 - \varphi_2)] = F_m \cdot \sin(\varphi_1 - \varphi_2) \cdot \sin(\varphi_3 - \varphi_2) \\ v_B^D \equiv v_{bn} = v_n \cdot \sin[\pi - (\varphi_3 - \varphi_2)] = v_m \cdot \sin(\varphi_1 - \varphi_2) \cdot \sin(\varphi_3 - \varphi_2) \end{cases} \\ \omega_3 = \frac{l_1 \cdot \sin(\varphi_1 - \varphi_2) \cdot \omega_1}{l_3 \cdot \sin(\varphi_3 - \varphi_2)} \Rightarrow v_B = l_3 \cdot \omega_3 = \frac{l_1 \cdot \omega_1 \cdot \sin(\varphi_1 - \varphi_2)}{\sin(\varphi_3 - \varphi_2)} = \frac{v_m \cdot \sin(\varphi_1 - \varphi_2)}{\sin(\varphi_3 - \varphi_2)} \\ v_B^D = D \cdot v_B \Leftrightarrow v_m \cdot \sin(\varphi_1 - \varphi_2) \cdot \sin(\varphi_3 - \varphi_2) = D \cdot \frac{v_m \cdot \sin(\varphi_1 - \varphi_2)}{\sin(\varphi_3 - \varphi_2)} \Rightarrow D = \sin^2(\varphi_3 - \varphi_2) \\ \eta_i = \frac{P_3}{P_1} = \frac{F_B \cdot v_B}{F_m \cdot v_m} = \frac{F_m \cdot \sin(\varphi_1 - \varphi_2) \cdot \sin(\varphi_3 - \varphi_2) \cdot \frac{v_m \cdot \sin(\varphi_1 - \varphi_2)}{\sin(\varphi_3 - \varphi_2)}}{F_m \cdot v_m} = \sin^2(\varphi_1 - \varphi_2) \\ \eta_i^D = \frac{P_3^D}{P_1} = \frac{F_B \cdot v_B^D}{F_m \cdot v_m} = \frac{F_m \cdot \sin(\varphi_1 - \varphi_2) \cdot \sin(\varphi_3 - \varphi_2) \cdot v_m \cdot \sin(\varphi_1 - \varphi_2) \cdot \sin(\varphi_3 - \varphi_2)}{F_m \cdot v_m} = \\ = \sin^2(\varphi_3 - \varphi_2) \cdot \sin^2(\varphi_1 - \varphi_2) = D \cdot \eta_i \end{cases} \quad (5)$$

4. Conclusions

The presented method is the most elegant and direct method to determine the kinematics planar quadrilateral mechanism.

Relationships used by this method allow and the determination of the dynamic system vibration. In the dynamic kinematics the constant rotation speed $\omega_1 = ct.$ gets a variable value $\omega_1^D = D \cdot \omega_1$.

References

[1] Pelecudi, Chr., ş.a., *Mecanisme*. E.D.P., Bucureşti, 1985.

CHAPTER XI

DETERMINING THE MECHANICAL EFFICIENCY OF OTTO ENGINE'S MECHANISM

ABSTRACT: *This chapter presents a few original elements about the dynamics and kinematics of piston mechanism, used like motor mechanism from OTTO engines. It presents an original method to determine the efficiency of the piston mechanism, used like motor mechanism. This method consists of eliminating the friction modulus. One determines the efficiency of the piston mechanism in two ways: 1. When the piston mechanism works like a motor; 2. When the piston mechanism works like a steam roller. Finally it determines the total motor efficiency, for the four cycle engine and for two cycle engine. With the relation of motor efficiency it optimizes the Otto mechanism, which is the principal mechanism from the internal-combustion engines. This is the way to diminish the acceleration of the piston and to maximize the efficiency of motor mechanism. It optimizes the constructive parameters: e, r, l, taking into account the rotation speed of drive shaft, n.*

Keywords: *Efficiency, force, piston, crank, connecting-rod, motor, stroke, bore.*

1. INTRODUCTION

In this chapter the authors present an original method to calculate the efficiency of the planar mechanisms with rods, concretely to the motor planar mechanism.

The originality consists of eliminating the friction forces and friction coefficients. It determines, directly the mechanical efficiency of planar rods mechanism only.

It determines the efficiency of piston mechanism (§ 2.) in two ways: 1. When the piston mechanism works like a motor; 2. When the piston mechanism works like a steam roller. Finally one determines the total motor efficiency, for the four cycle engine and for two cycle engine.

2. DETERMINING THE MECHANICAL MOTOR EFFICIENCY

In figure 1 it can see the kinematical diagram of the mechanism with crank - connecting rod - piston [1,2]. The constructive parameters are: r, the radius of crank; l, the length of connecting-rod; e, the eccentricity between centre of crank rotation and axis of piston guide. The mechanism is positioned by the angle, φ, which is representing the rotation angle of crank. The connecting rod is positioned by one of the two angles, α or ψ (see picture 1). The variable length between the centre of crank rotation and the piston centre is y_B.

2.1. The kinematics of Otto mechanism

The kinematical relations (see fig. 1) are the following:

$$\begin{cases} r \cdot \cos\varphi + l \cdot \cos\psi = -e \\ r \cdot \sin\varphi + l \cdot \sin\psi = y_B \end{cases} \quad (1)$$

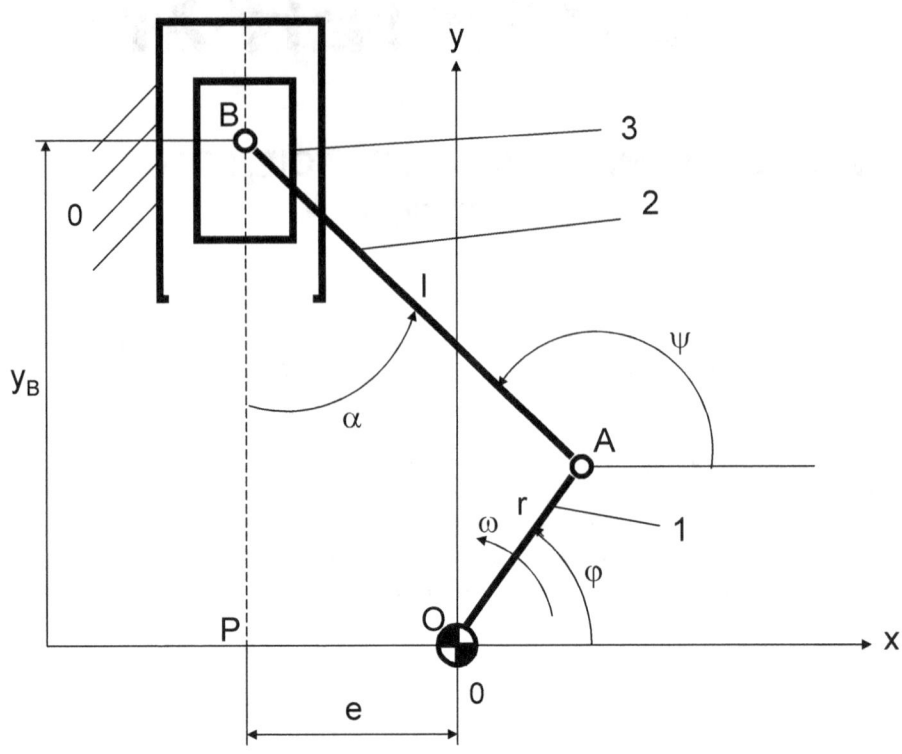

Fig. 1. *The kinematical schema of Otto-mechanism*

From first relation of the positions system (1) it determines the value of ψ angle (see the relation 2):

$$\cos\psi = -\frac{e + r\cdot\cos\varphi}{l} \tag{2}$$

From the second relation of system (1) it calculates directly the piston's displacement, $s=y_B$ (see the relation 3):

$$s = y_B = r\cdot\sin\varphi + l\cdot\sin\psi \tag{3}$$

We derivate the position's system (1) and obtains the velocity's system (4):

$$\begin{cases} -r\cdot\dot\varphi\cdot\sin\varphi - l\cdot\dot\psi\cdot\sin\psi = 0 \\ r\cdot\dot\varphi\cdot\cos\varphi + l\cdot\dot\psi\cdot\cos\psi = \dot y_B \end{cases} \tag{4}$$

From the first relation of system (4) it calculates the angular velocity, $\dot\psi$, (see the relation 5) and from the second relation of system (4) one determines the piston's linear velocity, $\dot y_B$, (see the relation 6):

$$\dot\psi = -\frac{r\cdot\sin\varphi}{l\cdot\sin\psi}\cdot\dot\varphi \tag{5}$$

$$\dot y_B = r\cdot\dot\varphi\cdot\cos\varphi + l\cdot\dot\psi\cdot\cos\psi \tag{6}$$

It derivates the velocity's system (4) and obtains the acceleration's system (7):

$$\begin{cases} -r\cdot\dot{\varphi}^2\cdot\cos\varphi - l\cdot\dot{\psi}^2\cdot\cos\psi - l\cdot\ddot{\psi}\cdot\sin\psi = 0 \\ -r\cdot\dot{\varphi}^2\cdot\sin\varphi - l\cdot\dot{\psi}^2\cdot\sin\psi + l\cdot\ddot{\psi}\cdot\cos\psi = \ddot{y}_B \end{cases} \quad (7)$$

From the first relation of system (7) it calculates the angular acceleration, $\ddot{\psi}$, (see the relation 8) and from the second relation of system (7) one determines the piston's linear acceleration, $\ddot{y}_B$, (relation 9):

$$\ddot{\psi} = -\frac{r\cdot\dot{\varphi}^2\cdot\cos\varphi + l\cdot\dot{\psi}^2\cdot\cos\psi}{l\cdot\sin\psi} \quad (8)$$

$$\ddot{y}_B = l\cdot\ddot{\psi}\cdot\cos\psi - r\cdot\dot{\varphi}^2\cdot\sin\varphi - l\cdot\dot{\psi}^2\cdot\sin\psi \quad (9)$$

The angle α can be put in a function of ψ angle, see the expression (10):

$$\alpha = \psi - 90 \quad (10)$$

From the system (11) it can now determine the trigonometric functions of the α angle:

$$\begin{cases} \cos\alpha = \sin\psi \\ \sin\alpha = -\cos\psi \end{cases} \quad (11)$$

With the expression (2) and the second relation of system (11), we can determine $\sin\alpha$, see the relation (12):

$$\sin\alpha = \frac{e + r\cdot\cos\varphi}{l} \quad (12)$$

The piston's velocity takes the form (13):

$$\begin{aligned} v_B = \dot{y}_B &= r\cdot\dot{\varphi}\cdot\cos\varphi + l\cdot\dot{\psi}\cdot\cos\psi = \\ &= r\cdot\dot{\varphi}\cdot\cos\varphi - \frac{r\cdot\dot{\varphi}\cdot\sin\varphi\cdot\cos\psi}{\sin\psi} = \\ &= \frac{r\cdot\dot{\varphi}}{\sin\psi}\cdot(\cos\varphi\cdot\sin\psi - \sin\varphi\cdot\cos\psi) = \\ &= r\cdot\dot{\varphi}\cdot\frac{\sin(\psi-\varphi)}{\sin\psi} = r\cdot\omega\cdot\frac{\sin(\psi-\varphi)}{\sin\psi} \end{aligned} \quad (13)$$

2.2. Determining the mechanical efficiency when the Otto mechanism works like a motor mechanism

The Otto mechanism works like a motor mechanism in a single cycle (π angle), when the piston is moving from the near dead point to the distant dead point (when the piston is moving from an extreme position to another – see the picture 2); this is the single motor time.

In the Figure 2 it can see the kinematical diagrams of Otto-mechanism in the extremely positions:

a) when the crank is in prolonging of the connecting-rod,

b) when the crank is overlapped on the connecting-rod

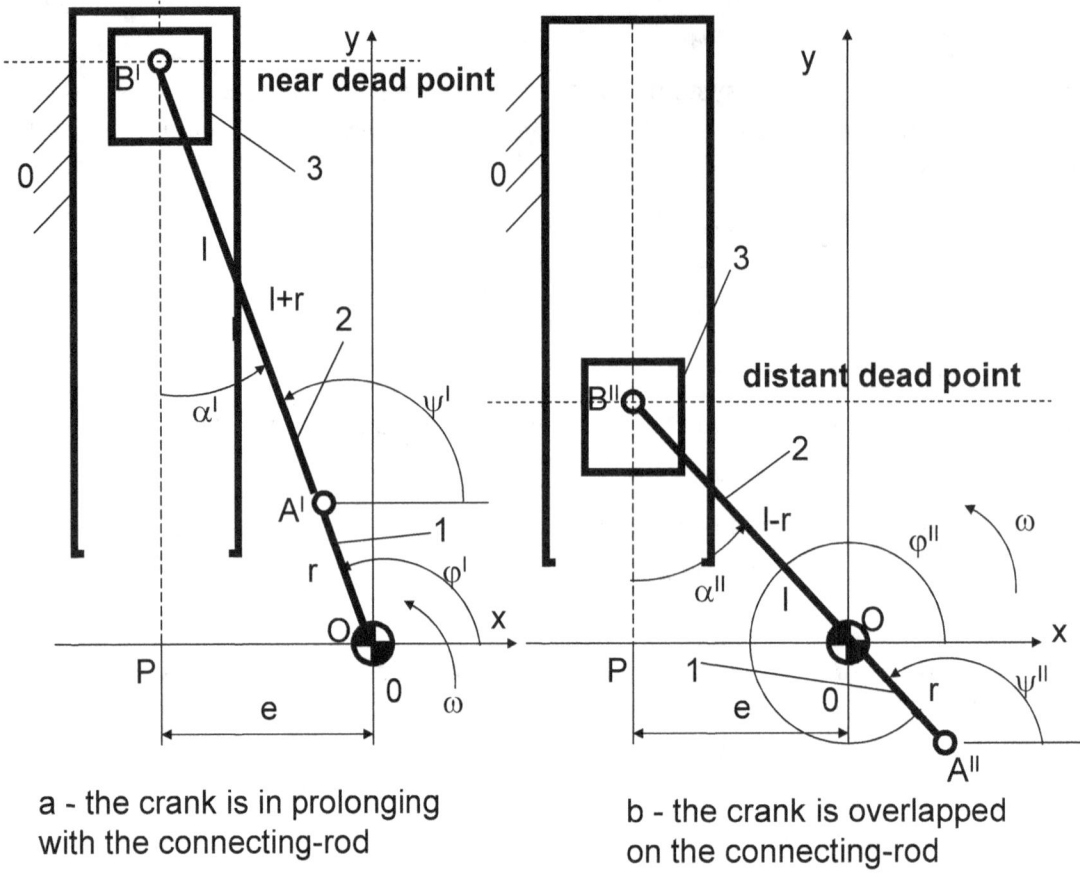

Fig. 2. *The kinematical diagrams of Otto-mechanism in the extremely positions; a) when the crank is in prolonging of the connecting-rod, b) when the crank is overlapped on the connecting-rod*

The efficiency of the pistons mechanism when the piston works like a motor mechanism can be determined, if we go from the piston to the crank, with the determining of forces (see the figure 3), [2,3].

The consumed motor force (the input force), F_m, is divided in two components: 1) F_n - the normal force (in the long of the connecting-rod); 2) F_τ - the tangential force (perpendicular in B, on the connecting-rod); see the system (14); (in figure 3 ω is negative; the rotation sense is hourly).

$$\begin{cases} F_n = F_m \cdot \cos\alpha = F_m \cdot \sin\psi \\ F_\tau = F_m \cdot \sin\alpha = -F_m \cdot \cos\psi \end{cases} \quad (14)$$

F_n is the single force transmitted from B to A (because the rod hasn't a joint with the frame).

In A, the force F_n is divided in two components too: 1. F_u – the utile force; 2. F_c – a compression force. See the system (15):

$$\begin{cases} F_u = F_n \cdot \sin(\psi - \varphi) = F_m \cdot \sin\psi \cdot \sin(\psi - \varphi) \\ F_c = F_n \cdot \cos(\psi - \varphi) = F_m \cdot \sin\psi \cdot \cos(\psi - \varphi) \end{cases} \quad (15)$$

The utile power, P_u, can be written in form (16):

$$\begin{aligned} P_u &= F_u \cdot v_A = F_u \cdot r \cdot \omega = \\ &= F_m \cdot r \cdot \omega \cdot \sin\psi \cdot \sin(\psi - \varphi) \end{aligned} \quad (16)$$

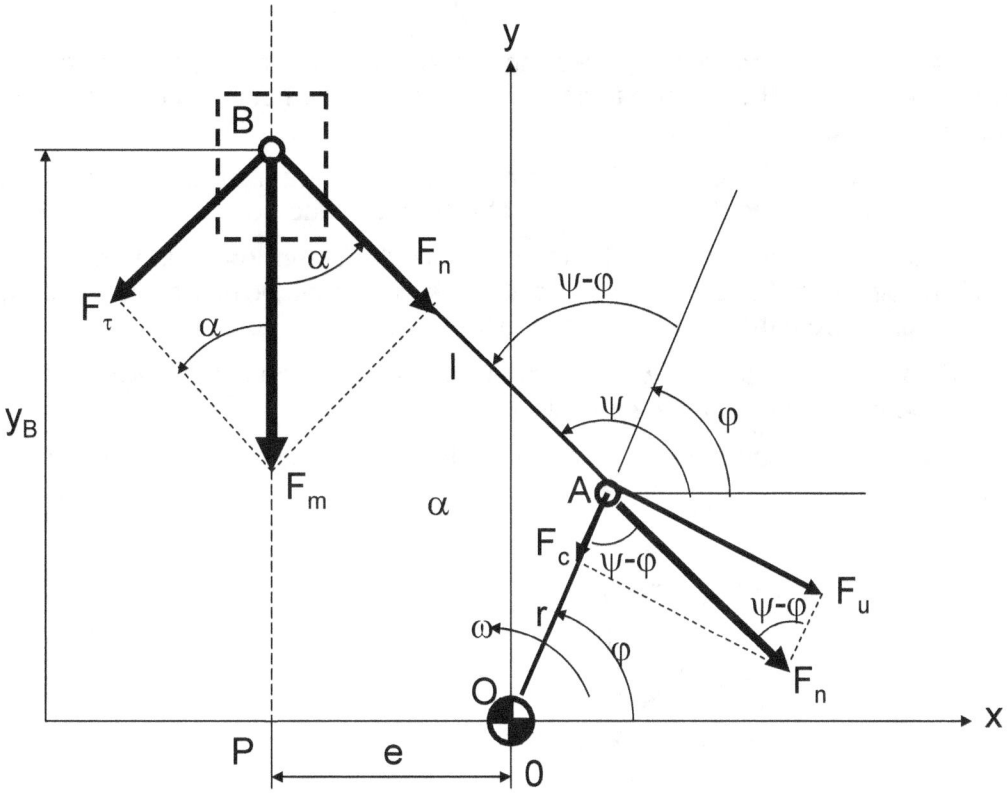

Fig. 3. *The forces of Otto-mechanism, when the piston works like a motor mechanism*

The consumed power, P_c, can be written in form (17):

$$P_c = F_m \cdot v_B = F_m \cdot r \cdot \omega \cdot \frac{\sin(\psi - \varphi)}{\sin \psi} \qquad (17)$$

The momentary mechanical efficiency, η_i, can be written with the relation (18):

$$\eta_i = \frac{P_u}{P_c} = \frac{F_m \cdot r \cdot \omega \cdot \sin \psi \cdot \sin(\psi - \varphi)}{F_m \cdot r \cdot \omega \cdot \sin(\psi - \varphi) \cdot \frac{1}{\sin \psi}} = \sin^2 \psi = \cos^2 \alpha = 1 - \frac{(e + r \cdot \cos \varphi)^2}{l^2} \qquad (18)$$

To calculate the mechanical efficiency, η, it can integrate the momentary efficiency, η_i, from near dead point to distant dead point, from φ^I to φ^{II} (figure 2):

$$\begin{cases} \varphi^I \equiv \varphi_i = \pi - a\cos(\frac{e}{l+r}) \\ \varphi^{II} \equiv \varphi_f = 2 \cdot \pi - a\cos(\frac{e}{l-r}) \end{cases} \qquad (19)$$

We determine approximately the efficiency with the relation (20), only if we can determine precisely the extreme angles, α_M and α_m:

$$\eta = 0.5 + \frac{\sin \alpha_M \cos \alpha_M - \sin \alpha_m \cos \alpha_m}{2 \cdot (\alpha_M - \alpha_m)} \qquad (20)$$

2.2. Determining the mechanical efficiency when the Otto mechanism works like steam roller

2.3.

The Otto mechanism works like motor mechanism in a single cycle (a π angle), when the piston is moving from the near dead point to the distant dead point, and it works like steam roller in the rest of the energetically cycle.

At the two cycle engines, the motor works like steam roller, in a single cycle, when the piston is moving from the distant dead point to the near dead point.

At the four cycle engines, the motor works like steam roller, in three cycle; two times the piston is moving from the distant dead point to the near dead point, and in one cycle (one time) the piston is moving from the near dead point to the distant dead point.

By a cycle (a π angle), we understand a time, a single time, precisely a semi kinematical-cycle; a kinematical cycle has a 2.π angle.

In figure 4 it can see the forces in Otto mechanism when the mechanism works like a steam roller.

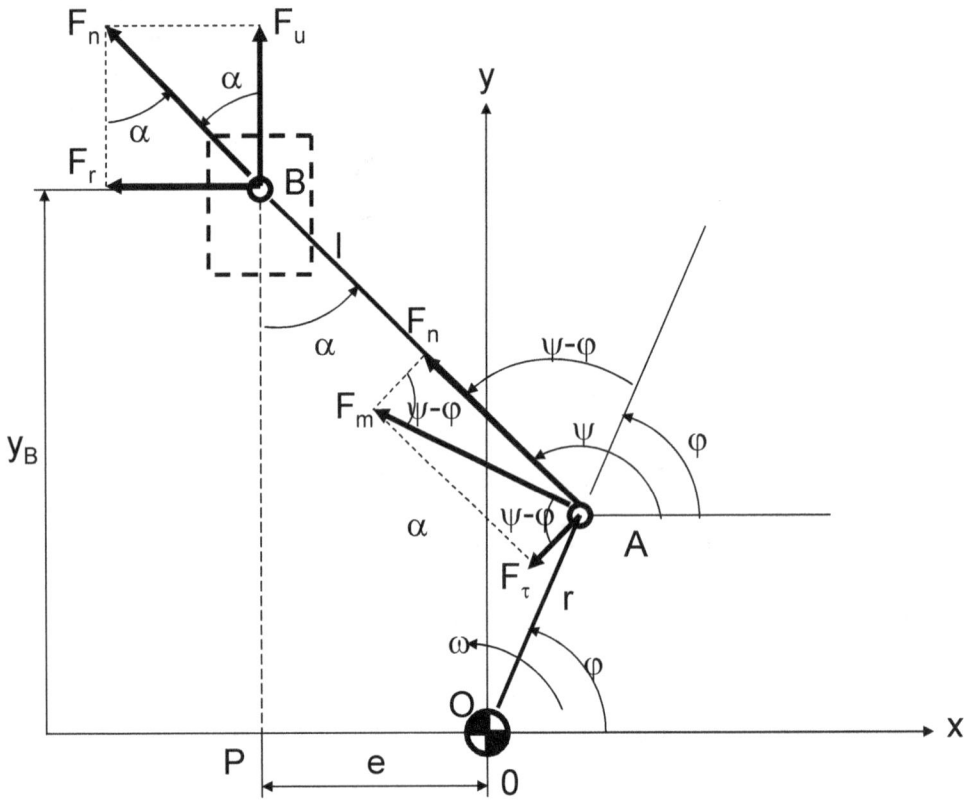

Fig. 4. *The forces of Otto-mechanism, when the piston works like a steam roller*

The input force (the consumed motor force), F_m, perpendicular in A on the crank OA (r), is divided in two components: 1. F_n–the normal force, which is the active component, the only components transmitted from couple A to joint B; 2. F_τ–the tangential force, which can give a couple, and can rotate the connecting-rod, or bend it, [2,3]; see the system (21):

$$\begin{cases} F_n = F_m \cdot \sin(\psi - \varphi) \\ F_\tau = F_m \cdot \cos(\psi - \varphi) \end{cases} \qquad (21)$$

In joint B, the transmitted force, F_n, is divided in two components too: 1. F_u – the useful force; 2. F_r – a force normal at the guide axis; see the system (22):

$$\begin{cases} F_u = F_n \cdot \cos\alpha = F_n \cdot \sin\psi = \\ \quad = F_m \cdot \sin(\psi-\varphi) \cdot \sin\psi \\ F_r = F_n \cdot \sin\alpha = -F_n \cdot \cos\psi = \\ \quad = -F_m \cdot \sin(\psi-\varphi) \cdot \cos\psi \end{cases} \quad (22)$$

The utile power can be written in form (23) and the consumed power can be written in form (24):

$$P_u = F_u \cdot v_B = F_m \cdot \sin(\psi-\varphi) \cdot \sin\psi \cdot \\ \cdot \frac{r\omega \sin(\psi-\varphi)}{\sin\psi} = F_m \cdot r \cdot \omega \cdot \sin^2(\psi-\varphi) \quad (23)$$

$$P_c = F_m \cdot v_A = F_m \cdot r \cdot \omega \quad (24)$$

The momentary mechanical efficiency when the piston works like steam roller, can be calculated with the relation (25):

$$\eta_i = \frac{P_u}{P_c} = \frac{F_m \cdot r \cdot \omega \cdot \sin^2(\psi-\varphi)}{F_m \cdot r \cdot \omega} = \sin^2(\psi-\varphi) = \\ = \frac{[\sqrt{l^2-(e+r\cdot\cos\varphi)^2} \cdot \cos\varphi + (e+r\cdot\cos\varphi)\cdot\sin\varphi]^2}{l^2} \quad (25)$$

3. CONCLUSIONS

The momentary mechanical efficiency when the piston works like steam roller (25), is different that the efficiency when the piston works like motor (18), (in the motor time). Generally the steam roller efficiency is lower that the motor efficiency.

The steam roller efficiency is approximately 50% or a lower value and the motor efficiency can be 60-99%, in function of the constructive parameters, e, r, l. The motor efficiency increases when the ratio, $\lambda=r/l$, decreases. For a $\lambda<0.33$, the motor efficiency is high enough.

The efficiency when the piston works like a steam roller may be calculated by integrating the momentary efficiency (25) in two ways: a) from φ^I to φ^{II}; b) from φ^{II} to φ^I. Generally the results are not the same. We must calculate three types of efficiency for the four cycle engines, and we should calculate two ways of efficiency for the two cycle engines.

The final efficiency for the four cycle engines can be 50-60%, and for the two cycle engines can be 73%, with e=10 [mm], r=20 [mm], l=90 [mm] for example (see figure 1). The two cycle engines can give us a 13% more efficiency (this fact may be important).

The Otto mechanism may be improved for giving a better efficiency and a minimum value for the maximum acceleration [3].

Constructive, we must adopt a lower stroke and a greater bore. The radius of crank, r, must be shorter. The piston should take the aspect of a pot (a frying pan; see [3]).

REFERENCES

[1] Pelecudi, Chr., ş.a., *Mecanisme*, E.D.P., Bucureşti, 1985;
[2] Petrescu, V., Petrescu, I., *Randamentul cuplei superioare de la angrenajele cu roţi dinţate cu axe fixe*, In: The Proceedings of 8th National Symposium PRASIC, Braşov, vol. I, pp. 333-338, 2002.
[3] Petrescu, F.I., Petrescu, R.V., *Câteva elemente privind îmbunătăţirea designului mecanismului motor*, In: The Proceedings of 8th National Symposium on GTD, Braşov, vol. I, pp. 353-358, 2003.

CHAPTER XII
OTTO ENGINE DYNAMICS

Abstract: *Otto engine dynamics are similar in almost all common internal combustion engines. We can speak so about dynamics of engines: Lenoir, Otto, and Diesel. The dynamic presented model is simple and original. The first thing necessary in the calculation of Otto engine dynamics, is to determine the inertial mass reduced at the piston. It uses then the Lagrange equation.*

Key words: *Lagrange equation, dynamic model, inertial mass, Otto dynamics*

1. INTRODUCTION

The first thing necessary in the calculation of Otto engine dynamics, is to determine the inertial mass reduced at the piston (1).

$$\begin{cases} M \equiv M^* = m_t + m_{bA} \cdot \dfrac{r^2}{s'^2} + \dfrac{J_1}{s'^2} + \dfrac{J_2}{s'^2} \cdot \dfrac{\lambda^2 \cdot \cos^2 \varphi}{\cos^2 \alpha} \\ \\ M = m_t + [(m_{bA} + \dfrac{J_1}{r^2}) \cdot (1 - \lambda^2 \cdot \sin^2 \varphi) + \dfrac{J_2}{l^2} \cdot \cos^2 \varphi] \cdot \\ \quad \cdot \dfrac{1}{\sin^2 \varphi \cdot (\cos \alpha + \lambda \cdot \cos \varphi)^2} \\ \\ M = m_t + \dfrac{m_1 \cdot (1 - \lambda^2 \cdot \sin^2 \varphi) + m_2 \cdot \cos^2 \varphi}{\sin^2 \varphi \cdot (\cos \alpha + \lambda \cdot \cos \varphi)^2} \end{cases} \quad (1)$$

Then it derives the reduced mass to the crank position angle (2). Were used for piston the next kinematics parameters (4). Lagrange equation is written in the form (3).

$$\begin{aligned} \dfrac{dM}{d\varphi} &= (M - m_t) \cdot (-2) \cdot (\dfrac{\cos \varphi}{\sin \varphi} - \dfrac{\lambda \cdot \sin \varphi}{\cos \alpha}) - \\ &\quad - \dfrac{2 \cdot \cos \varphi \cdot (\lambda^2 \cdot m_1 + m_2)}{\sin \varphi \cdot (\cos \alpha + \lambda \cdot \cos \varphi)^2} \end{aligned} \quad (2)$$

$$M \cdot \omega^2 \cdot x'' + \dfrac{1}{2} \cdot \dfrac{dM}{d\varphi} \cdot \omega^2 \cdot x' = k \cdot (s - x) - F_p \quad (3)$$

$$\begin{cases} s = r \cdot \cos \varphi + l \cdot \cos \alpha - l \\ s' = -\dfrac{r \cdot \sin \varphi}{\cos \alpha} \cdot (\cos \alpha + \lambda \cdot \cos \varphi) \\ s'' = -r \cdot \cos \varphi - \dfrac{r \cdot \lambda \cdot \cos(2\varphi)}{\cos \alpha} - \dfrac{r \cdot \lambda^3 \cdot \sin^2 \varphi \cdot \cos^2 \varphi}{\cos^3 \alpha} \end{cases} \quad (4)$$

2. DYNAMIC EQUATIONS

The dynamic equation of motion of the piston, obtained by integrating the Lagrange equation (3), takes the form 5.

$$x = s \cdot \sqrt[3]{\frac{k}{k - m_t \cdot \omega^2}} - c_3 \cdot \frac{\cos\varphi}{\cos\alpha \cdot (\cos\alpha + \lambda \cdot \cos\varphi)} + c_4 \cdot \cos\varphi \qquad (5)$$

Dynamic reduced velocity (6) and dynamic reduced acceleration (7) are obtained by derivation:

$$x' = s' \cdot \sqrt[3]{\frac{k}{k - m_t \cdot \omega^2}} + c_3 \cdot \frac{\sin\varphi}{\cos\alpha \cdot (\cos\alpha + \lambda \cdot \cos\varphi)} - c_4 \cdot \sin\varphi \qquad (6)$$

$$x'' = s'' \cdot \sqrt[3]{\frac{k}{k - m_t \cdot \omega^2}} + c_3 \cdot \frac{\cos\varphi}{\cos\alpha \cdot (\cos\alpha + \lambda \cdot \cos\varphi)} - c_4 \cdot \cos\varphi \qquad (7)$$

Angular velocity ω^* is obtained through kinetic energy conservation (8-12).

$$\frac{1}{2} \cdot J^* \cdot \omega^{*2} = \frac{1}{2} \cdot J_D^* \cdot \omega_D^2 \qquad (8)$$

$$\begin{aligned}\omega_D &= \omega_m \cdot D = \omega_m \cdot (\cos\alpha)^2 = \omega_m \cdot \cos^2\alpha = \\ &= \omega_m \cdot (1 - \sin^2\alpha) = \omega_m \cdot (1 - \lambda^2 \cdot \sin^2\varphi)\end{aligned} \qquad (9)$$

$$J^* = J_1 + m_{bA} \cdot r^2 + m_t \cdot s'^2 \qquad (10)$$

$$J_D^* = J_1 + m_{bA} \cdot r^2 + m_t \cdot x'^2 \qquad (11)$$

$$\omega^* = \sqrt{\frac{J_1 + m_{bA} \cdot r^2 + m_t \cdot x'^2}{J_1 + m_{bA} \cdot r^2 + m_t \cdot s'^2}} \cdot \frac{\pi \cdot n}{30} \cdot (1 - \lambda^2 \cdot \sin^2\varphi) \qquad (12)$$

Dynamic velocity (13) and kinematics velocity (14) are written:

$$\dot{x} = x' \cdot \omega^* \qquad (13)$$

$$\dot{s} = s' \cdot \omega_m = s' \cdot \frac{\pi \cdot n}{30} \qquad (14)$$

Dynamic acceleration (15) and kinematics acceleration (16) are written:

$$\ddot{x} = x'' \cdot \omega^{*2} \tag{15}$$

$$\ddot{s} = s'' \cdot \omega_m^2 = s'' \cdot \frac{\pi^2 \cdot n^2}{900} \tag{16}$$

3. NOTATIONS

In the picture number 1 it presents the crank shaft.

$$k = \frac{3 \cdot \pi \cdot E \cdot G \cdot (D_m^4 - d_m^4)}{4G(l_m + b)^3 + 96Er^2 \sin^2 \varphi (D_m^4 - d_m^4)[\frac{l_p + .4D_p}{D_p^4 - d_p^4} + \frac{l_m + .4D_m}{D_m^4 - d_m^4} + \frac{8r - 1.6(D_p + D_m)}{b(2r + D_p + D_m)^3}]} \tag{17}$$

The relation (17) determines the elastic constant of the crank shaft, k. For the masses it uses the notations (18); see the picture two.

$\lambda \Rightarrow$ the ratio between lengths of crank and rod; $\lambda = \dfrac{r}{l}$

$m_p \Rightarrow$ the mass of the piston, with piston bolt and segments;

$m_b \Rightarrow$ the mass of the rod;

$$\begin{cases} m_{bA} = m_b \cdot \dfrac{l''}{l} \quad m_{bB} = m_b \cdot \dfrac{l'}{l} \quad l'+l''=l \quad m_{bA} + m_{bB} = m_b \\ m_t = m_p + m_{bB} \\ m_1 = m_{bA} + \dfrac{J_1}{r^2} \\ m_2 = \dfrac{J_2}{l^2} \end{cases} \tag{18}$$

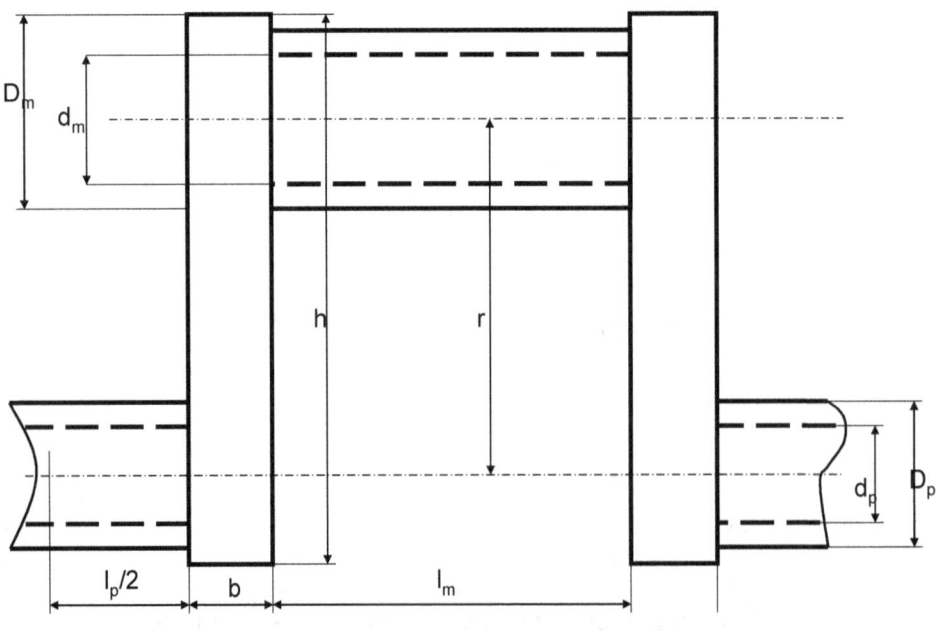

Fig. 1 *Crank Shaft*

The parameters c1-c4 take the forms (19):

$$\begin{cases} c_1 = \dfrac{r}{k} \cdot \omega^2 \quad [\dfrac{m}{kg}] \\ \\ c_2 = \lambda^2 \cdot m_1 + m_2 \quad [kg] \\ \\ c_3 = c_1 \cdot c_2 \quad [m] \\ \\ c_4 = c_1 \cdot m_t \quad [m] \end{cases} \quad (19)$$

The moment of inertia J_1 can be determined with the relation (20).

$$J_1 = \frac{\pi \cdot \rho}{32} \cdot \{(l_p + 2 \cdot b) \cdot (D_p^4 - d_p^4) + \\ (l_m + 2 \cdot b) \cdot [(D_m^4 - d_m^4) + (D_m^2 - d_m^2) \cdot 8 \cdot r^2]\} \quad (20)$$

The crank length, r, and the length of the connecting-rod, l, can be seen in the kinematics schema of an Otto mechanism (fig. 2).

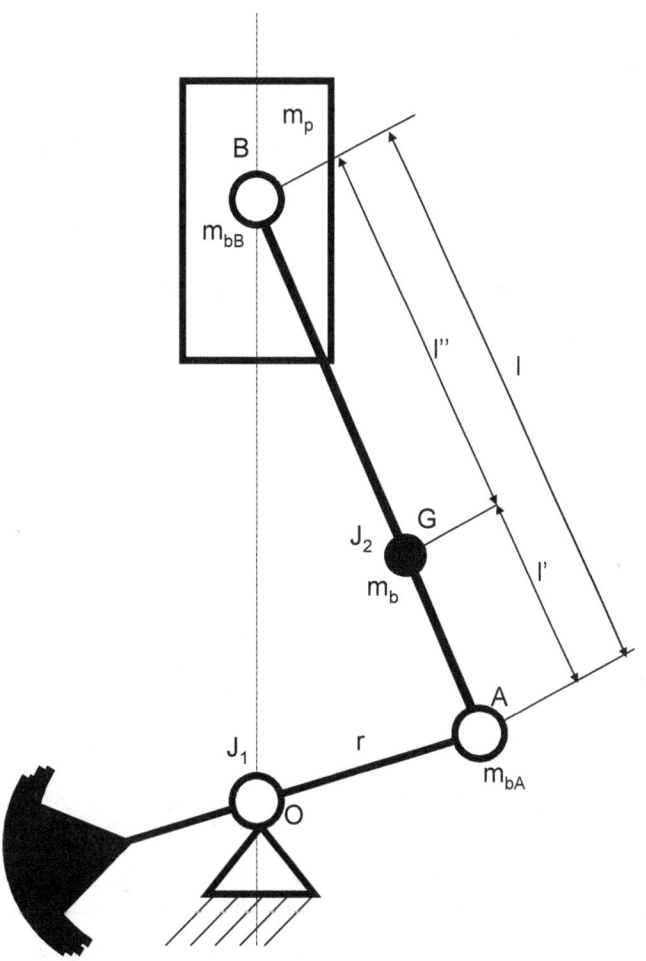

Fig. 2 *Otto mechanism kinematics schema*

4. DYNAMIC ANALYSIS OF THE MECHANISM AND CONCLUSIONS

When λ increases the mechanism dynamics is deteriorating.
r=0.25 [m] l=0.3 [m] $\lambda = 0.8(3)$

For n=8000 [r/m] the mechanism is working normally (see the accelerations diagram from the picture 3):

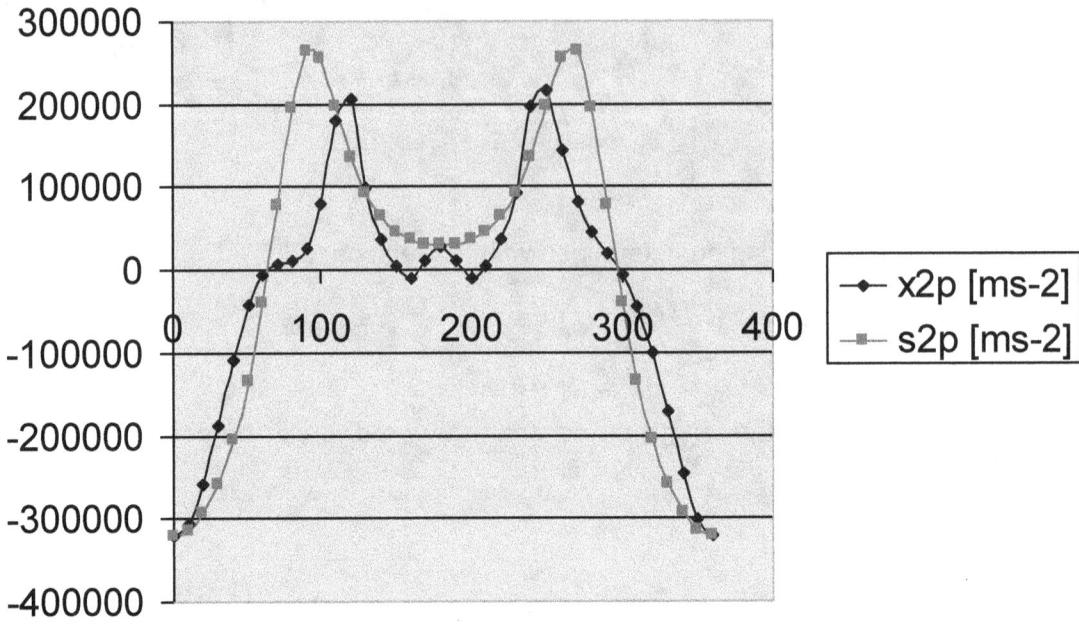

Fig. 3 Dynamic and kinematics accelerations; n=8000 [r/m]; $\lambda = 0.83$ r=0.25 [m] l=0.3 [m] $\lambda = 0.8(3)$

At n=9000 [r/m] the mechanism work abnormally (see the accelerations diagram from the picture 4):

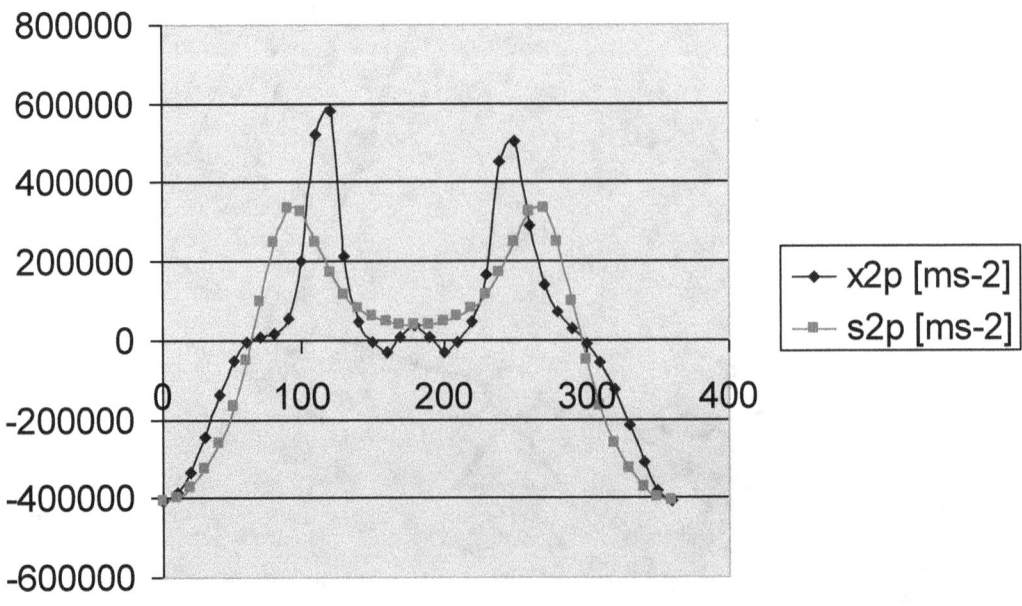

Fig. 4 Dynamic and kinematics accelerations; n=9000 [r/m]; $\lambda = 0.83$ r=0.25[m];l=0.3[m]

For a proper operation is necessary reduction of the ratio λ, especially if we want to increase the engine speed (see the next diagrams).

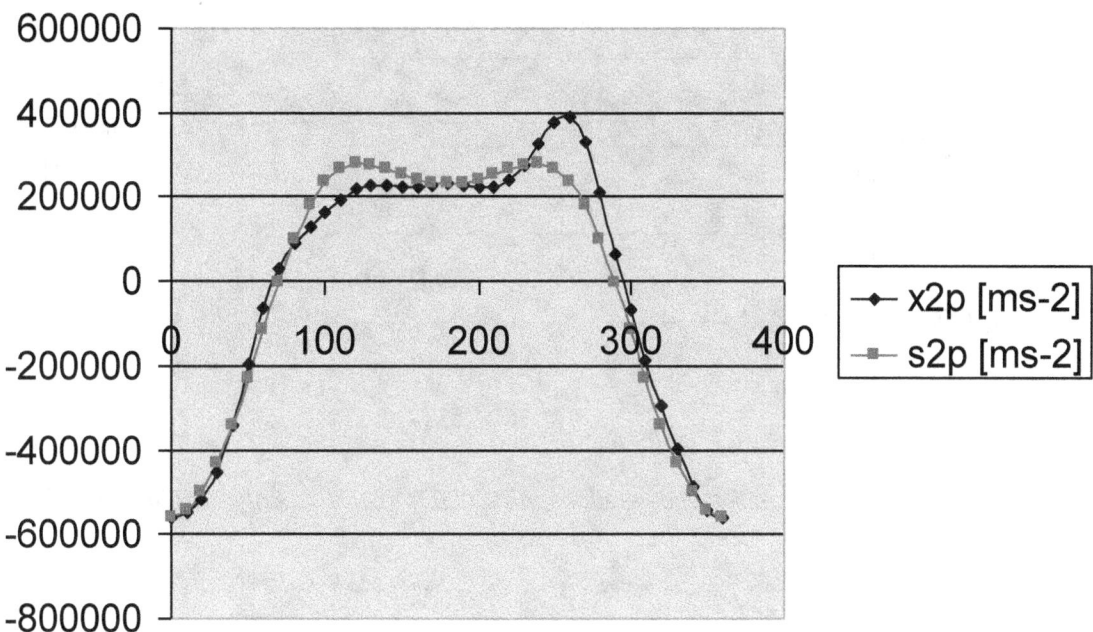

Fig. 5 *Dynamic and kinematics accelerations; n=12000 [r/m]; r=0.25[m];l=0.6[m]* $\lambda = 0.42$

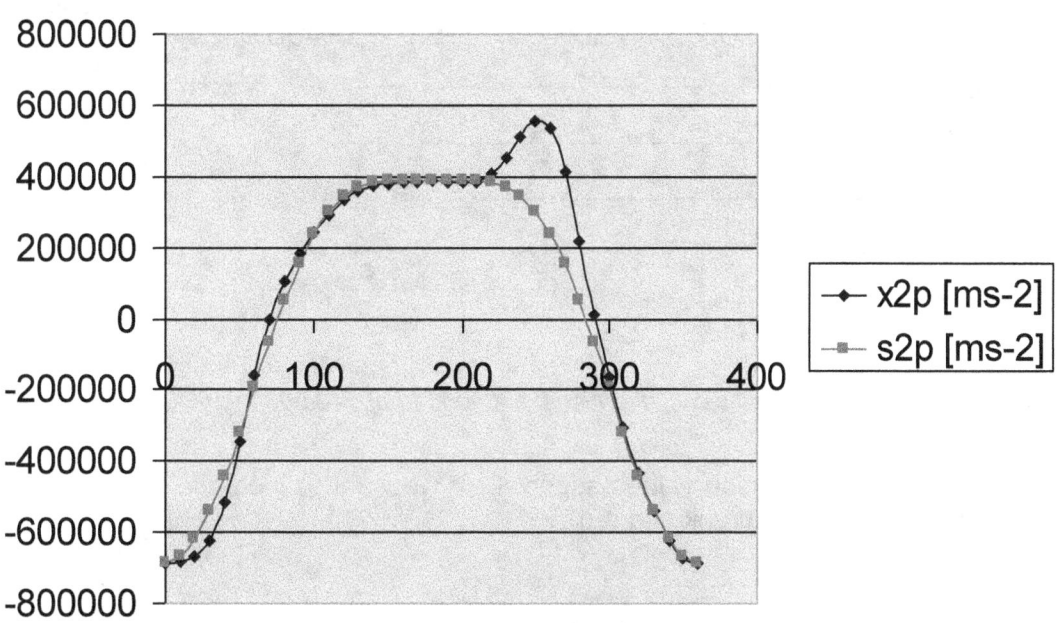

Fig. 6 *Dynamic and kinematics accelerations; n=14000 [r/m]; r=0.25[m];l=0.9[m]* $\lambda = 0.27$

We can reduce the acceleration values by reducing r and l.

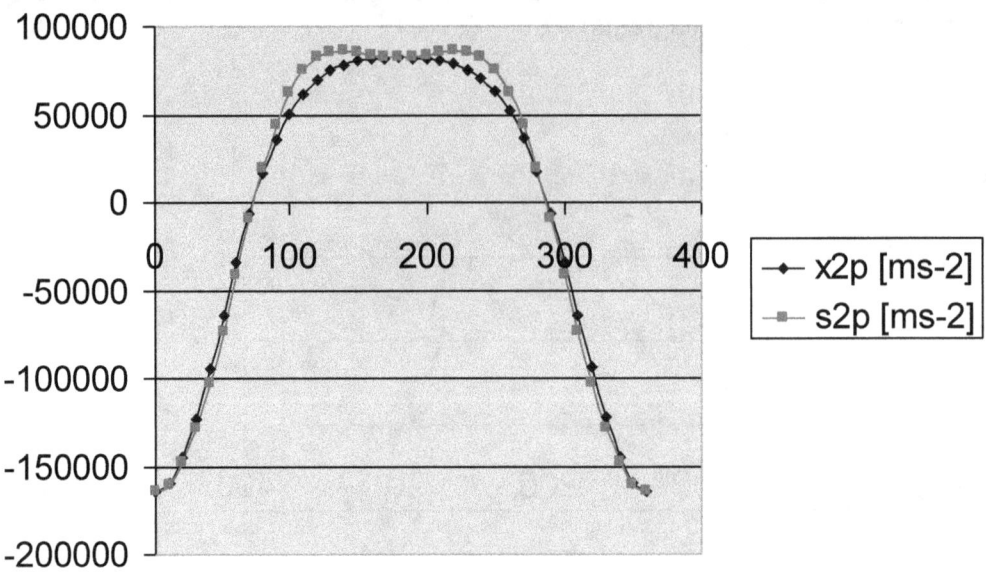

Fig. 7 *Dynamic and kinematics accelerations; n=15000 [r/m]; r=0.05[m];l=0.15[m]* $\lambda = 0.33$

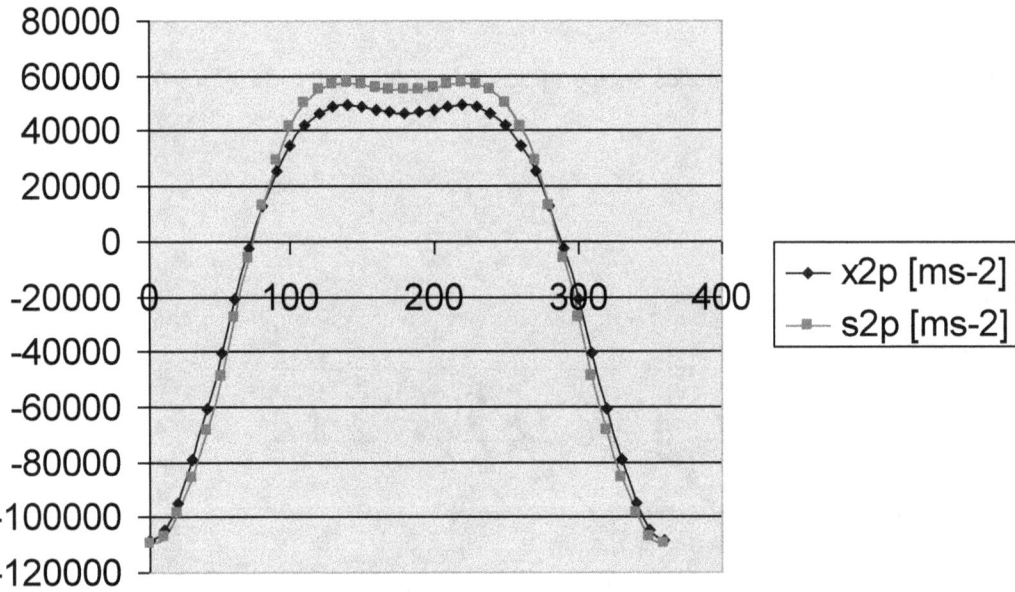

Fig. 8 *Dynamic and kinematics accelerations; n=50000 [r/m]; r=0.003[m];l=0.009[m]* $\lambda = 0.33$

It can reduce the acceleration values especially if we want to increase the engine speed by reducing r and l (the lengths of crank and rod).

5. REFERENCES

[1] Petrescu, R.V., Petrescu, F.I., - Otto Engine Design, Acta Technica Napocensis, Series: Applied Mathematics and Mechanics, CNCSIS 118 B, ISSN 1221-5872, Vol. Ib, p. 537-540, Cluj-Napoca, 2009
[2] Petrescu, F.I., Petrescu, R.V., - V Engine Design, Acta Technica Napocensis, Series: Applied Mathematics and Mechanics, CNCSIS 118 B, ISSN 1221-5872, Vol. Ib, p. 533-536, Cluj-Napoca, 2009

CHAPTER XIII
AN ORIGINAL INTERNAL COMBUSTION ENGINE

ABSTRACT: *The chapter presents a new and original internal-combustion engine. It is presenting a method in determining the kinematics and the efficiency of a new mechanism, MF1, proposed (designed) to work and be tested like an internal-combustion engine. It determines the mechanical momentary efficiency, when the mechanism works like a steam roller and when the mechanism works like a motor. The determined efficiency is different in the two described situations. It presents an original way to determine the dynamic efficiency as well. The dynamic momentary efficiency is the same in the two situations: when the mechanism works like a steam roller and when it works like a motor. We can determine the efficiency without friction, but we may anytime add the effect of friction modulus. It presents the dynamic kinematics of this mechanism too: the dynamic velocity and the dynamic acceleration. When the constructive parameters are normal, the dynamic velocities take the same values like the classical speeds and the dynamic accelerations take the same values like the classical accelerations.*

Keywords: Motor, connecting-rod, dynamic-velocity, dynamic-acceleration, dynamic-efficiency.

1. INTRODUCTION

The chapter shortly presents a new and original internal-combustion engine.
The originality consists in the way of determining the mechanical and dynamic efficiency and in the way of determining the dynamic velocities and accelerations.

2. PRESENTING THE KINEMATICS OF MF1

In picture number 1, we can see the kinematics outline of the mechanism of the new presented motor (Motor Florio 1), [2,3].

The first modification of this model, having in view the classical model (Otto engine mechanism), is the use of two connecting-rod, (2 and 3) and the use of B couple, a dual couple: of rotation and translation.

This motor mechanism is a new mechanism and his functionality will be different from the classical mechanism's functionality. The great advantage of this mechanism is that it can be regulated to have a bigger zone with constant acceleration at the piston (the element number five). The efficiency of this mechanism is the same like the classical Otto mechanism. The structural group 2-4 (a dyad) can improve the motor functionality without damage of power. The kinematics relations are the following (1-11):

$$a^2 = l_0^2 + l_1^2 - 2 \cdot l_0 \cdot l_1 \cdot \sin\varphi_1 \qquad (1)$$

$$\cos\varphi_2 = -\frac{l_1 \cdot \cos\varphi_1}{a} \qquad (2)$$

$$\cos\varphi_3 = \frac{e - l_1 \cdot \cos\varphi_1 - l_2 \cdot \cos\varphi_2}{l_3} \qquad (3)$$

$$y_D = l_1 \cdot \sin\varphi_1 + l_2 \cdot \sin\varphi_2 + l_3 \cdot \sin\varphi_3 \qquad (4)$$

$$\omega_2 = -\frac{l_1 \cdot \cos(\varphi_1 - \varphi_2)}{a} \cdot \omega_1 \qquad (5)$$

$$\dot{a} = -\frac{l_0 \cdot l_1 \cdot \omega_1 \cdot \cos\varphi_1}{a} \tag{6}$$

MOTOR FLORIO1-MF1
© 2003 Florian PETRESCU
The Copyright-Law
Of March, 01, 1989
U.S. Copyright Office
Library of Congress
Washington, DC 20559-6000
202-707-3000

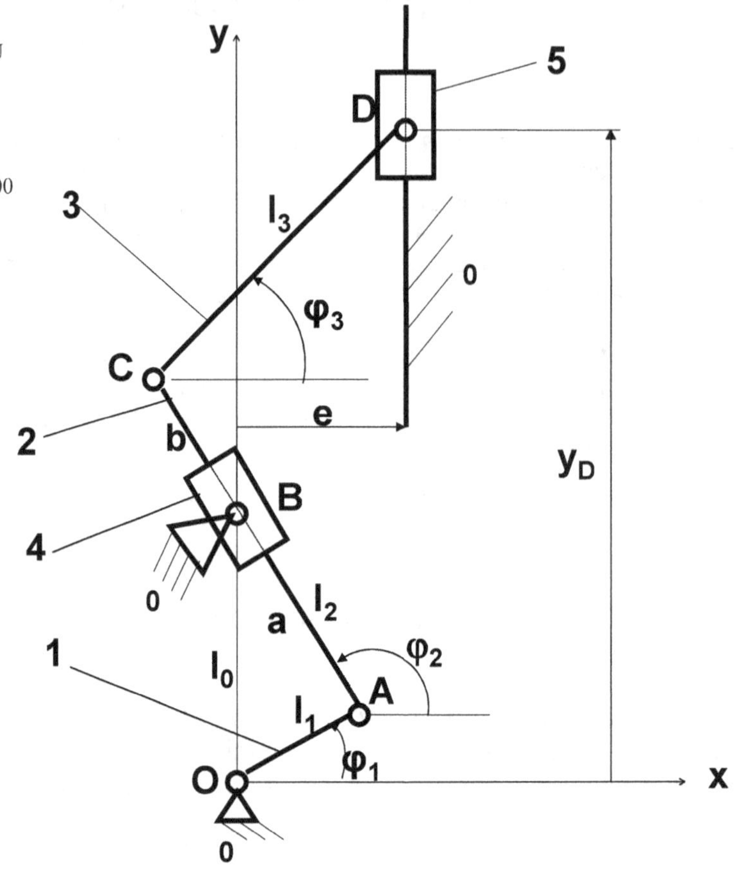

Fig. 1. *The MF1 kinematics outline*

$$\omega_3 = \frac{l_1 \cdot \omega_1}{a \cdot l_3} \cdot [l_0 \cdot \cos\varphi_1 \cdot \sin(\varphi_3 - \varphi_2) + b \cdot \cos(\varphi_1 - \varphi_2) \cdot \cos(\varphi_3 - \varphi_2)] \tag{7}$$

$$\dot{y}_D = \omega_1 \cdot l_1 \cdot \cos\varphi_1 + \omega_2 \cdot l_2 \cdot \cos\varphi_2 + \omega_3 \cdot l_3 \cdot \cos\varphi_3 \tag{8}$$

$$\varepsilon_2 = \frac{l_1 \cdot \omega_1 \cdot (\omega_1 - \omega_2) \cdot \sin(\varphi_1 - \varphi_2) - \omega_2 \cdot \dot{a}}{a} \tag{9}$$

$$\begin{aligned}\varepsilon_3 = \frac{l_1 \cdot \omega_1}{a \cdot l_3} \cdot [&-l_0 \cdot \omega_1 \cdot \sin\varphi_1 \cdot \sin(\varphi_3 - \varphi_2) + \\ &+ l_0 \cdot (\omega_3 - \omega_2) \cdot \cos\varphi_1 \cdot \cos(\varphi_3 - \varphi_2) - \\ &b \cdot (\omega_1 - \omega_2) \cdot \sin(\varphi_1 - \varphi_2) \cdot \cos(\varphi_3 - \varphi_2) - \\ &- b \cdot (\omega_3 - \omega_2) \cdot \cos(\varphi_1 - \varphi_2) \cdot \sin(\varphi_3 - \varphi_2)] - \frac{\dot{a}}{a} \cdot \omega_3\end{aligned} \tag{10}$$

$$\begin{aligned}\ddot{y}_D = &-\omega_1^2 \cdot l_1 \cdot \sin\varphi_1 - \omega_2^2 \cdot l_2 \cdot \sin\varphi_2 + \\ &+ \varepsilon_2 \cdot l_2 \cdot \cos\varphi_2 - \omega_3^2 \cdot l_3 \cdot \sin\varphi_3 + \varepsilon_3 \cdot l_3 \cdot \cos\varphi_3\end{aligned} \tag{11}$$

1. DETERMINING THE MOMENTARY MECHANICAL EFFICIENCY WHEN THE MECHANISM WORKS LIKE A STEAM ROLLER

It can determine the momentary mechanical efficiency, when the mechanism works like a steam roller, if one determines the distribution of forces, from the crank to the piston (figure 2); relations (12-19) [2,3]:

$$\begin{cases} F_n = F_m \cdot \sin(\varphi_2 - \varphi_1) \\ F_{\tau_A} = F_m \cdot \cos(\varphi_2 - \varphi_1) \end{cases} \quad (12)$$

$$F_{\tau_C} = \frac{a}{b} \cdot F_{\tau_A} = \frac{a}{b} \cdot F_m \cdot \cos(\varphi_2 - \varphi_1) \quad (13)$$

$$\begin{cases} F_n^I = F_n \cdot \cos(\varphi_2 - \varphi_3) \\ F_{\tau_C}^I = F_{\tau_C} \cdot \sin(\varphi_2 - \varphi_3) \end{cases} \quad (14)$$

$$\begin{aligned} F_T &= F_n^I + F_{\tau_C}^I = F_n \cdot \cos(\varphi_2 - \varphi_3) + F_{\tau_C} \cdot \sin(\varphi_2 - \varphi_3) = \\ &= F_m \cdot \sin(\varphi_2 - \varphi_1) \cdot \cos(\varphi_2 - \varphi_3) + \frac{a}{b} \cdot F_m \cdot \cos(\varphi_2 - \varphi_1) \cdot \sin(\varphi_2 - \varphi_3) = \\ &= F_m \cdot [\sin(\varphi_2 - \varphi_1) \cdot \cos(\varphi_2 - \varphi_3) + \frac{a}{b} \cdot \cos(\varphi_2 - \varphi_1) \cdot \sin(\varphi_2 - \varphi_3)] \end{aligned} \quad (15)$$

$$\begin{cases} F_U = F_T \cdot \sin \varphi_3 \\ F_R = F_T \cdot \cos \varphi_3 \end{cases} \quad (16)$$

$$F_U = F_m \cdot \sin \varphi_3 \cdot [\sin(\varphi_2 - \varphi_1) \cdot \cos(\varphi_2 - \varphi_3) + \\ + \frac{a}{b} \cdot \cos(\varphi_2 - \varphi_1) \cdot \sin(\varphi_2 - \varphi_3)] \quad (17)$$

$$v_U = v_m \cdot [\cos \varphi_1 - \frac{l_2 \cdot \cos \varphi_2 \cdot \cos(\varphi_1 - \varphi_2)}{a} + \\ \frac{l_0 \cdot \cos \varphi_1 \cdot \sin(\varphi_3 - \varphi_2) + b \cdot \cos(\varphi_1 - \varphi_2) \cdot \cos(\varphi_3 - \varphi_2)}{a} \cdot \cos \varphi_3] \quad (18)$$

$$\eta_{iC} = \frac{F_U \cdot v_U}{F_m \cdot v_m} = \sin \varphi_3 \cdot [\sin(\varphi_2 - \varphi_1) \cdot \cos(\varphi_2 - \varphi_3) + \\ + \frac{a}{b} \cdot \cos(\varphi_2 - \varphi_1) \cdot \sin(\varphi_2 - \varphi_3)] \cdot \\ \cdot [\cos \varphi_1 - \frac{l_2 \cdot \cos \varphi_2 \cdot \cos(\varphi_1 - \varphi_2)}{a} + \\ + \frac{l_0 \cdot \cos \varphi_1 \cdot \sin(\varphi_3 - \varphi_2) + b \cdot \cos(\varphi_1 - \varphi_2) \cdot \cos(\varphi_3 - \varphi_2)}{a} \cdot \cos \varphi_3] \quad (19)$$

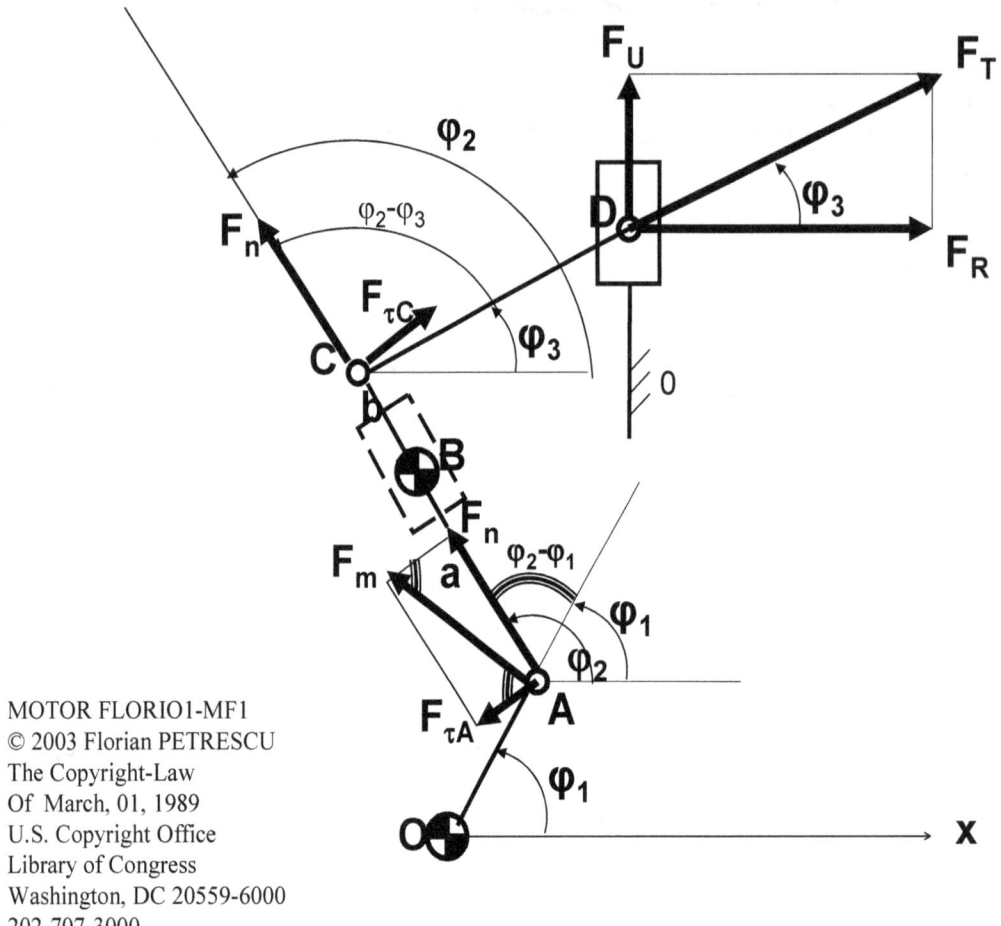

Fig. 2. *The MF1 distribution of forces, when the mechanism works like a steam roller*

4. DETERMINING THE MOMENTARY MECHANICAL EFFICIENCY WHEN THE MECHANISM WORKS LIKE A MOTOR

It can determine the momentary mechanical efficiency, when the mechanism works like a motor, if one determines the distribution of forces, from the piston to the crank (figure 3); relations (20-25) [2,3]:

$$\begin{cases} F_N = F_m \cdot \sin\varphi_3 \\ F_R = F_m \cdot \cos\varphi_3 \end{cases} \quad (20)$$

$$\begin{cases} F_n = F_N \cdot \cos(\varphi_2 - \varphi_3) \\ F_{\tau_C} = F_N \cdot \sin(\varphi_2 - \varphi_3) \end{cases} \quad (21)$$

$$F_{\tau_A} = \frac{b}{a} \cdot F_{\tau_C} = \frac{b}{a} \cdot F_m \cdot \sin\varphi_3 \cdot \sin(\varphi_2 - \varphi_3) \quad (22)$$

$$\begin{cases} F_{u1} = F_n \cdot \sin(\varphi_1 - \varphi_2) \\ F_{u2} = -F_{\tau_A} \cdot \cos(\varphi_1 - \varphi_2) \end{cases} \quad (23)$$

$$F_u = F_{u1} + F_{u2} = F_m \cdot \sin\varphi_3 \cdot [\cos(\varphi_2 - \varphi_3) \cdot \sin(\varphi_1 - \varphi_2) - \frac{b}{a} \cdot \sin(\varphi_2 - \varphi_3) \cdot \cos(\varphi_1 - \varphi_2)] \quad (24)$$

$$\mu_{iM} = \frac{-\sin\varphi_3 \cdot [\cos(\varphi_2 - \varphi_3) \cdot \sin(\varphi_1 - \varphi_2) - \dfrac{b}{a} \cdot \sin(\varphi_2 - \varphi_3) \cdot \cos(\varphi_1 - \varphi_2)]}{\cos\varphi_1 + \cos\varphi_3 \dfrac{l_0 \cos\varphi_1 \sin(\varphi_3 - \varphi_2) + b\cos(\varphi_1 - \varphi_2)\cos(\varphi_3 - \varphi_2)}{a} - \dfrac{l_2 \cos\varphi_2 \cos(\varphi_1 - \varphi_2)}{a}} \quad (25)$$

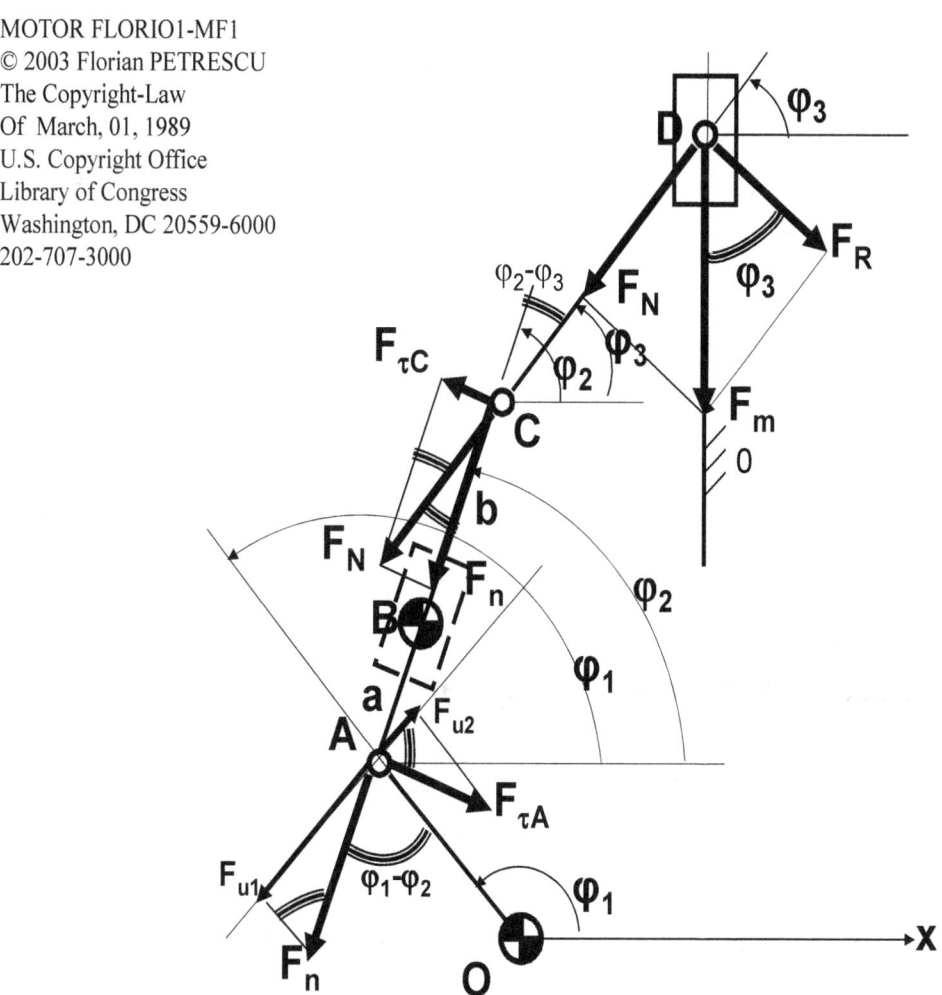

MOTOR FLORIO1-MF1
© 2003 Florian PETRESCU
The Copyright-Law
Of March, 01, 1989
U.S. Copyright Office
Library of Congress
Washington, DC 20559-6000
202-707-3000

Fig. 3. *The MF1 distribution of forces, when the mechanism works like a motor*

5. DETERMINING THE MOMENTARY DYNAMIC EFFICIENCY

The dynamic efficiency of the mechanism is the same, anytime (when the mechanism works like a steam roller and when it's working like a motor). It can be determined approximately with the relation (26):

$$\begin{cases} \mu_i^D = \mu_{iM}^D = \eta_{iC}^D = \sin^2\varphi_3 \cdot \sin^2\tau \\ with: \quad \tau = 2\cdot\varphi_2 - \varphi_1 - \varphi_3 \end{cases} \qquad (26)$$

It can determine the exactly momentary dynamic efficiency of the mechanism, if one takes in calculation the dynamic velocities (in this case the speeds distribution is the same like the forces distribution), see the relations (27-29):

$$-\frac{F_u}{F_m} = \sin\varphi_3 \cdot [\sin(\varphi_2 - \varphi_1)\cdot\cos(\varphi_2 - \varphi_3) + \frac{b}{a}\cdot\sin(\varphi_2 - \varphi_3)\cdot\cos(\varphi_2 - \varphi_1)] \qquad (27)$$

$$-\frac{v_u}{v_m} = \sin\varphi_3 \cdot [\sin(\varphi_2 - \varphi_1)\cdot\cos(\varphi_2 - \varphi_3) + \frac{a}{b}\cdot\sin(\varphi_2 - \varphi_3)\cdot\cos(\varphi_2 - \varphi_1)] \qquad (28)$$

$$\mu_i^D = \sin^2\varphi_3 \cdot \{\sin^2(\varphi_2 - \varphi_1)\cdot\cos^2(\varphi_2 - \varphi_3) + \sin^2(\varphi_2 - \varphi_3)\cdot\cos^2(\varphi_2 - \varphi_1) + \frac{a^2 + b^2}{4\cdot a\cdot b}\cdot\sin[2\cdot(\varphi_2 - \varphi_1)]\cdot\sin[2\cdot(\varphi_2 - \varphi_3)]\} \qquad (29)$$

6. THE DYNAMIC KINEMATICS OF THE MECHANISM

We can determine now the dynamic velocity (30) and the dynamic acceleration of the piston (31):

$$v_D^{Din} = l_1 \cdot \omega_1 \cdot \sin\varphi_3 \cdot [\sin(\varphi_2 - \varphi_1)\cdot\cos(\varphi_2 - \varphi_3) + \frac{a}{b}\cdot\sin(\varphi_2 - \varphi_3)\cdot\cos(\varphi_2 - \varphi_1)] \qquad (30)$$

$$a_D^{Din} = \{\omega_3 \cdot \cos\varphi_3 \cdot [\sin(\varphi_2 - \varphi_1)\cdot\cos(\varphi_2 - \varphi_3) + \frac{a}{b}\cdot\cos(\varphi_2 - \varphi_1)\cdot\sin(\varphi_2 - \varphi_3)] + \\ + \sin\varphi_3 \cdot [\cos(\varphi_2 - \varphi_1)\cdot\cos(\varphi_2 - \varphi_3)\cdot(\omega_2 - \omega_1) - \sin(\varphi_2 - \varphi_1)\cdot\sin(\varphi_2 - \varphi_3)\cdot(\omega_2 - \omega_3) - \\ - \frac{a}{b}\cdot\sin(\varphi_2 - \varphi_1)\cdot\sin(\varphi_2 - \varphi_3)\cdot(\omega_2 - \omega_1) + \frac{a}{b}\cdot\cos(\varphi_2 - \varphi_1)\cdot\cos(\varphi_2 - \varphi_3)\cdot(\omega_2 - \omega_3) + \\ + \frac{\dot{a}}{b}\cdot\cos(\varphi_2 - \varphi_1)\cdot\sin(\varphi_2 - \varphi_3) + \frac{a\cdot\dot{a}}{b^2}\cdot\cos(\varphi_2 - \varphi_1)\cdot\sin(\varphi_2 - \varphi_3)]\}\cdot l_1\cdot\omega_1 \qquad (31)$$

7. DISCUTION

If the values of the constructive parameters of the mechanism are normal, the dynamic speeds and the dynamic acceleration of the piston (30-31), are practical the same like the classical kinematics values (8-11), see the picture number (4, 5 and 6):

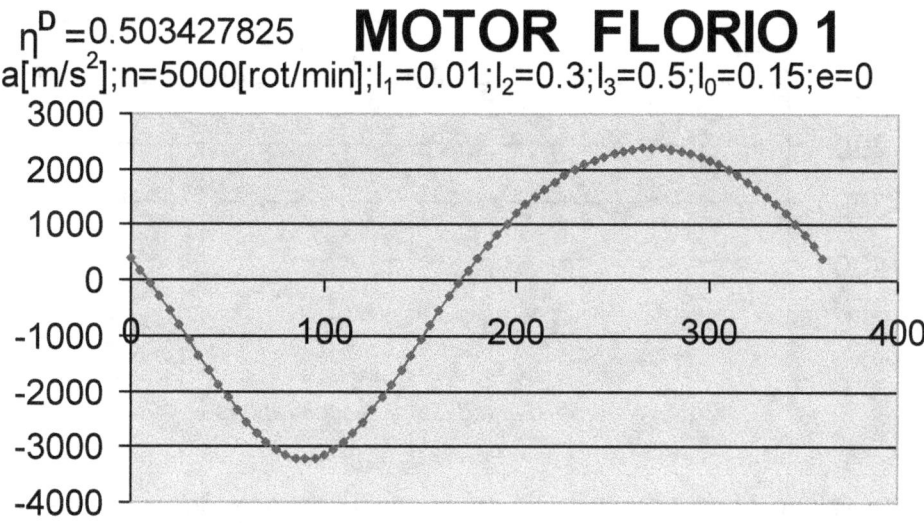

Fig. 4. *The MF1 piston acceleration, when the constructive parameters are normal*

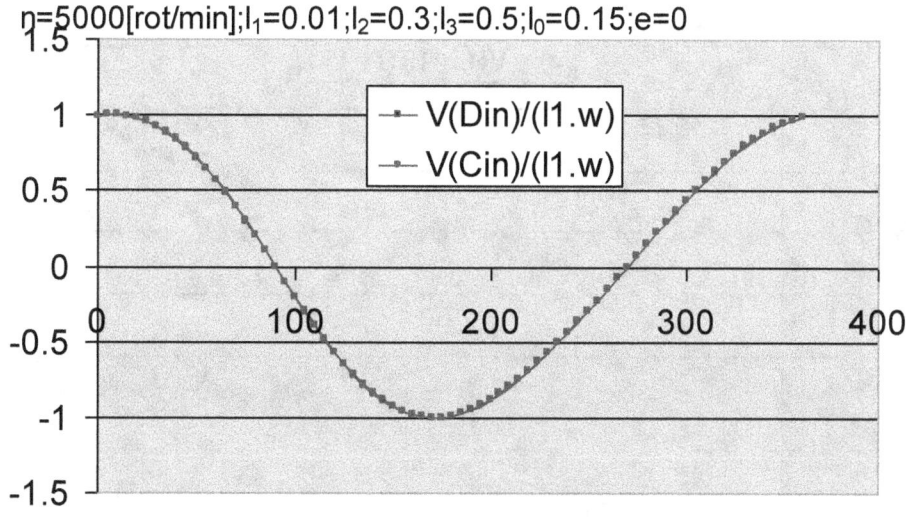

Fig. 5. *Kinematic and dynamic velocitie*

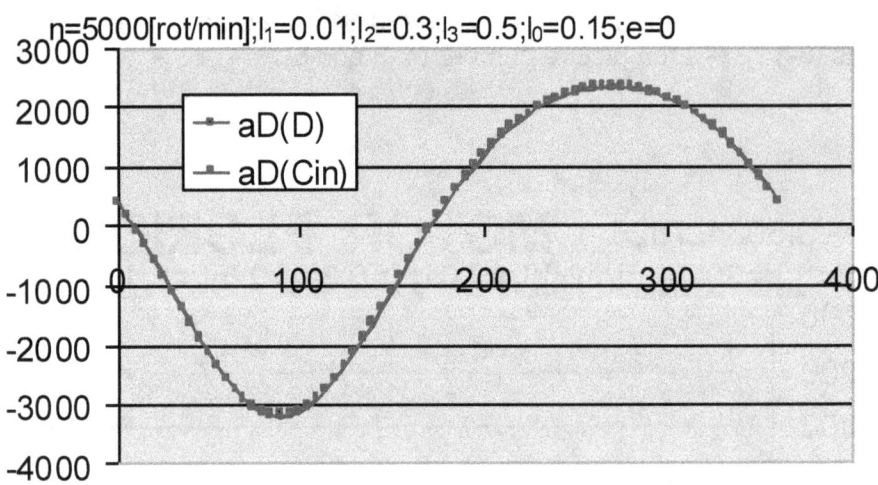

Fig. 6. *Kinematic and dynamic accelerations*

When the values of the constructive parameters are different from the normal classic parameters, the dynamic speeds and the dynamic acceleration of the piston (relations 30-31), are not the same like the classical kinematics values (8-11), see the pictures number (7 and 8):

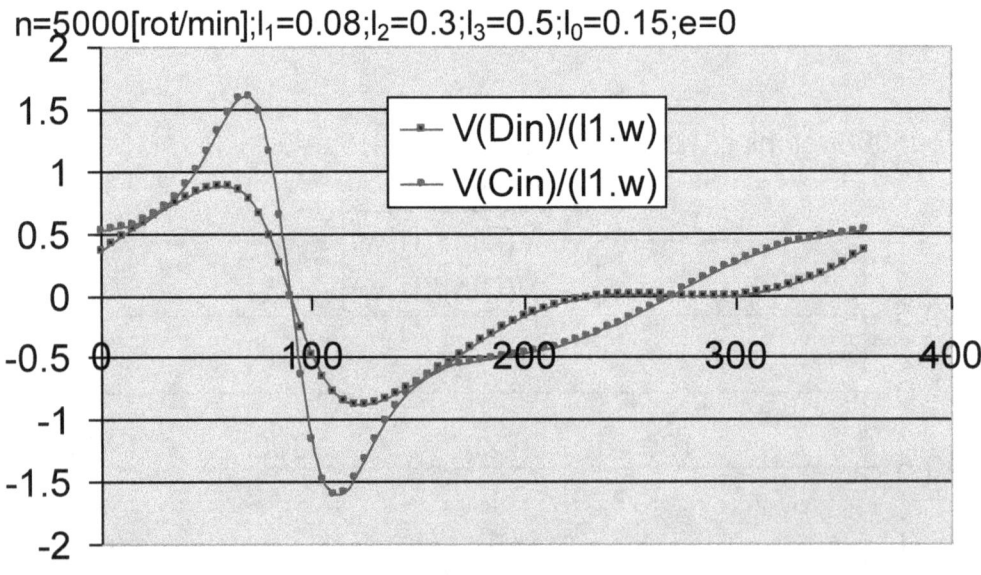

Fig. 7. *Kinematic and dynamic velocities*

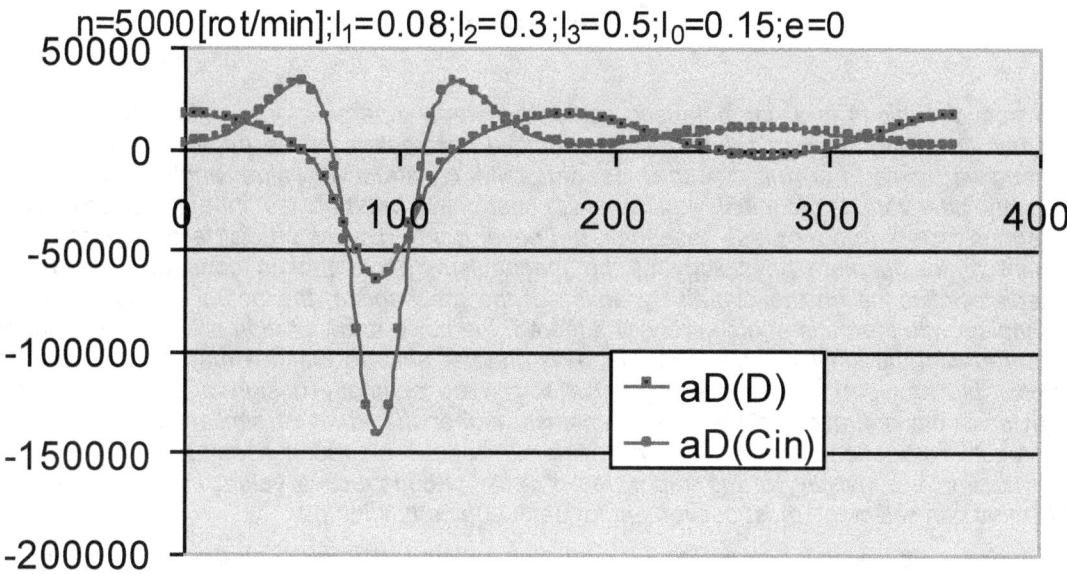

Fig. 8. *Kinematic and dynamic accelerations*

8. CONCLUSION

Some mechanisms have the same parameters for the classical and for the dynamic kinematics (gears, cams with plate followers, the planar tetra-later mechanism, etc...); Others don't. At the presented mechanism, the dynamic-kinematics is different from the classical-kinematics, but, if the constructive parameters are normal, the dynamic velocities practically take the same values like the classical speeds and the dynamic accelerations take the same values like the classical accelerations. Structurally, the mechanism has two dyad, when it works like a steam roller and it generates a triad, when works like motor.

REFERENCES

[1] Pelecudi, Chr., ş.a., *Mecanisme.* E.D.P., Bucureşti, 1985.
[2] Petrescu, V., Petrescu, I., *Randamentul cuplei superioare de la angrenajele cu roţi dinţate cu axe fixe,* In: The Proceedings of 7th National Symposium PRASIC, Braşov, vol. I, pp. 333-338, 2002.
[3] Petrescu, F.I., Petrescu, R.V., *Câteva elemente privind îmbunătăţirea designului mecanismului motor,* In: The Proceedings of 8th National Symposium on GTD, Braşov, vol. I, pp. 353-358, 2003.

CHAPTER XIV
V Engine Design

Abstract: *V engines in a characteristic aside, their reply kinematics-dynamic (operating in a dynamic viewpoint) is closely linked to constructive parameters of the engine, especially the constructive angle. For this reason, as generally constructive value angle was chosen randomly, after various technical requirements constructive or otherwise, inherited or calculated by various factors (more or less essential), but never got to discuss crucial factor (which takes account of the intimate physiology of the mechanism) angle that is constructive with his immediate influence on the overall dynamics of the mechanism, the actual dynamics of the mechanism with the main engine in the V suffered, the noise and vibration are generally higher compared with the similar engines in line. This chapter aims to make a major contribution to remedy this problem so that the engine in V can be optimally designed and its dynamic behavior in the operation to become blameless, higher than that of similar engines in line. Theoretical calculations are difficult and complex, but the alteration constructive required of them is simple, consisting of the imposition of a list of constructive values of the angle from which you can select the most convenient for each engine builder in V.*

Key words: *Efficiency, force, piston, crank, connecting-rod, motor, stroke, bore, dynamic-velocity, dynamic-acceleration, dynamic-efficiency, V engines, kinematic.*

1. INTRODUCTION

Kinematics and dynamic synthesis of V engines can be done according to the constructive alpha angle (α). This constructive angle alpha (see Figure 1) was elected in generally follow different criteria or design requirements (it is determined by the number of cylinders and the condition for obtaining the ignitions uniformly distributed).

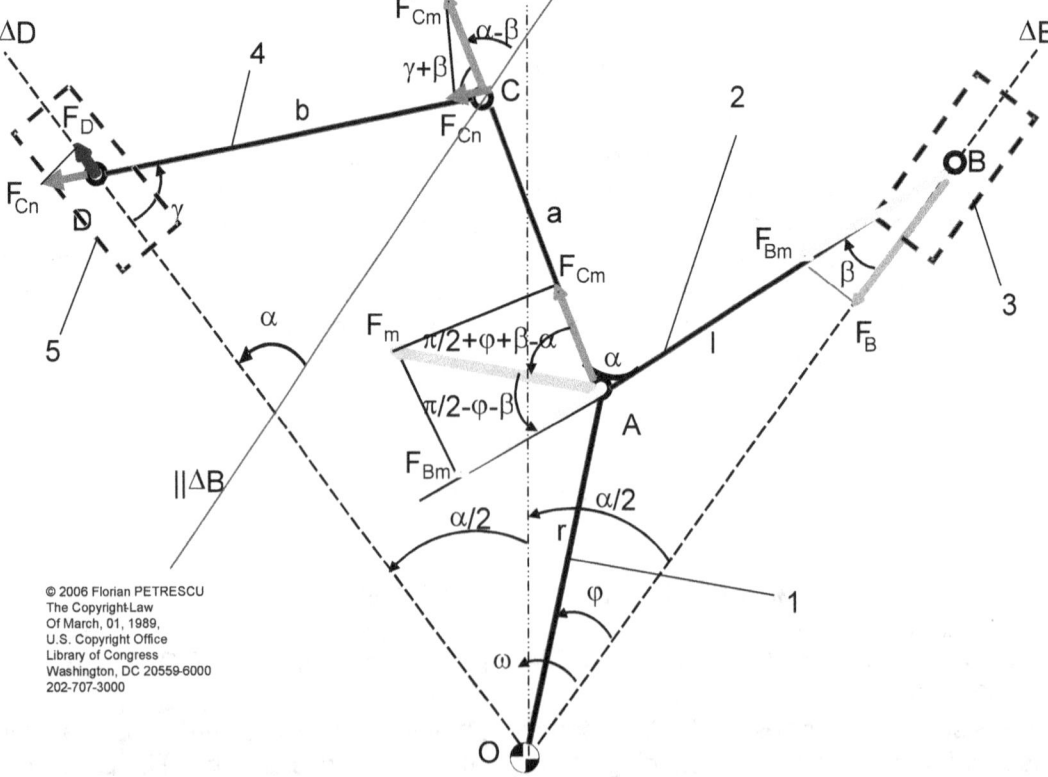

V Motors' Kinematics and Dynamics Synthesis by the Constructive Angle Value (α); Forces Distribution, Angles, Elements and Couples (Joints) Positions; a+b=l

Fig. 1. *Forces and Kinematics of V Engine*

This work proposes aggregating this angle after rigorous kinematics-dynamic criteria, so that the V engine obtained works silently, with vibration and noise much lower. It is even mainly disadvantage of a motor in V namely that it works with higher vibrations compared with a same power engine in line [1, 6-12].

The authors have studied this works for several years together with a collective joint research (UPB-Autobuzul plant) and dynamic behavior of the V engines [6-8], the level of vibration and noise produced, the level of vibration transmitted inside the vehicle, the possibility of limiting them through various solutions of gripping and containment of the engine.

Results were good but not very good. After similar measurements done on other types of engines it was decided the use of engines in line, much quieter than the V. Meanwhile engines have improved but also international standards that limit vibration and noise levels have become more demanding.

The engine in V has many fans; it is more compact, more dynamic, more robust, stronger, and higher operating efficiency compared to similar engines in line. But its fans are not only racing fans, motor and habit, there are a wide audience of consumers who want only cars equipped with nerve in V. As to conciliate them well and those who make rules to limit emissions of cars, it was thought that chapter aims to provide an equitable solution regarding engines in V.

2. THE BASIC IDEA

After decades of work in the mechanisms and machinery field, through experience, I noticed an interesting fact.

At the engines in line forces and velocities transmission is normal from the driver (motor) shaft (from the crank shaft) to the pistons (through the rods), and vice versa (in the engine times).

The engine in V transmission forces and velocities between elements is forced and unequal regardless of the meaning of transmission (from crank to pistons or from pistons to crank).

Dynamics imposed to the main piston is one, and the secondary piston is required another, so that dynamic speeds (actual speeds required) differ, and with them and pistons to crank feedback (to crank shaft), as each would require another main shaft speed. If that's a pair of pistons, for more pairs of pistons jerk resultant operation will be more and more powerful.

Obvious solution is to optimize the dynamic of each pair of pistons in hand. This optimization was based on dynamic coefficients of each piston.

Dynamic coefficient of a piston showing the actual crank angular velocity varies compared to required average angular velocity imposed by the motor shaft rotation speed.

This variation [3, 4] is due to several factors, kinematics and dynamic, being itself a function and of engine constructive parameters.

The usual mechanisms have a single dynamic factor (coefficient), as is the case and in-line engines.

At the engine in V appear two dynamic factors imposed to the crank (and to the crank shaft) by the two pistons linked together (secondary piston rod link to the main piston rod; see the Figure 1). The two dynamic factors differ among themselves and changes their values permanently depending on the crank angle position (crank shaft position).

This indicates that each piston (the primary and secondary) tries to impose its dynamic to the main shaft, so that the end result is an operation to struggle, since the two sets of pistons shoot one at one side and the other somewhere else.

The possible solution (unique solution) is matching the two dynamic factors.

More specifically to write a mathematical relationship to match the main piston (main engine) dynamic coefficient expression with that of the secondary piston (secondary motor) (Now you can see that engine in V is built of two engines merged, see the fig. 1). Relations that results are quite complicated [5].

Optimization based on obtained relationship can be made in several ways.

The most natural seems to be the optimization parameters in view of the engine builders of the V, particularly based on constructive angle alpha, which appears twice in the cinematic scheme of an engine in V: first it is mounting angle formed by the two axes of the two pistons coupled (angle formed by the axis of the main piston guide axis with the secondary piston guide axis); and the second time this item (constructive angle) appears on the element 2 (the rod of the main piston) between the two arms of the element 2 (AB and AC).

3. THE ENGINE SYNTHESIS
3.1. Presentation

In the picture number 1 we can see the kinematics schema of the V Engine. The crank 1 has a trigonometric rotation (ω) and actions the connecting-rod 2 which moves the piston 3 along the slide bar ΔB and actions the second connecting-rod 4, which moves the second piston 5 along the slide bar ΔD. There is a constructive angle α between the two axes ΔB and ΔD.

The same constructive angle (α) is formed by the two arms of the connecting-rod 2; first arm has the length l, and the second (which transmits the movement to the second connecting-rod 4) has the length a; this length a, add with the length b of the second connecting-rod 4 must gives the length l of the first connecting-rod.

The crank motor force F_m is perpendicular at the crank length r, in A.

A part of it (F_{Bm}) is transmitted to the first arm of connecting-rod 2 (along l) towards the first piston 3. Another part of the motor force, (F_{Cm}) is transmitted towards the second piston 5, by (along) the second arm of first connecting-rod 2 (a).

$$F_{B_m} = x \cdot F_m \cdot \cos[\frac{\pi}{2} - (\varphi + \beta)] = x \cdot F_m \cdot \sin(\varphi + \beta) \qquad (1)$$

$$F_{C_m} = y \cdot F_m \cdot \cos[\frac{\pi}{2} + \varphi + \beta - \alpha] = y \cdot F_m \cdot \sin(\alpha - \varphi - \beta) \qquad (2)$$

3.2. Forces and velocities

A percent (of motor force F_m) x is transmitted towards the first piston (element 3) and the percent y is transmitted towards the second piston (element 5); the sum between x and y is 1 or 100%.

The dynamic velocities have the same direction like forces:

$$v_{B_m} = x \cdot v_m \cdot \cos[\frac{\pi}{2} - (\varphi + \beta)] = x \cdot v_m \cdot \sin(\varphi + \beta) \qquad (3)$$

$$v_{C_m} = y \cdot v_m \cdot \cos[\frac{\pi}{2} + \varphi + \beta - \alpha] = y \cdot v_m \cdot \sin(\alpha - \varphi - \beta) \qquad (4)$$

From the element 2 (first arm) to the first piston (element 3) one transmits the force F_B (5) and the dynamic velocity v_{BD} (6).

$$F_B = F_{B_m} \cdot \cos\beta = x \cdot F_m \cdot \sin(\varphi + \beta) \cdot \cos\beta \qquad (5)$$

$$v_{B_D} = v_{B_m} \cdot \cos\beta = x \cdot v_m \cdot \sin(\varphi + \beta) \cdot \cos\beta \qquad (6)$$

The kinematics (known) velocity (imposed by the linkage) is given by the relation 7.

$$v_B = v_m \cdot \sin(\varphi + \beta) \cdot \frac{1}{\cos\beta} \qquad (7)$$

To force the first piston velocity equalises the dynamic value, it introduces a dynamic coefficient D_B (8):

$$D_B = x \cdot \cos^2 \beta \qquad (8)$$

Where,

$$v_{B_D} = D_B \cdot v_B \qquad (9)$$

$$v_m = r \cdot \omega \qquad (10)$$

The second Motor' outline can be solved now. In C, F_{Cm} and v_{Cm} are projected in F_{Cn} and v_{Cn}:

$$F_{C_n} = F_{C_m} \cdot \cos(\gamma + \beta) = y \cdot F_m \cdot \sin(\alpha - \varphi - \beta) \cdot \cos(\gamma + \beta) \qquad (11)$$

$$v_{C_n} = v_{C_m} \cdot \cos(\gamma + \beta) = y \cdot v_m \cdot \sin(\alpha - \varphi - \beta) \cdot \cos(\gamma + \beta) \qquad (12)$$

The transmitted force along of the second connecting-rod (F_{Cn}) is projected in D on the ΔD axe in F_D:

$$F_D = F_{C_n} \cdot \cos\gamma = y \cdot F_m \cdot \sin(\alpha - \varphi - \beta) \cdot \cos(\gamma + \beta) \cdot \cos\gamma \qquad (13)$$

The dynamic velocity in D is:

$$v_D = v_{C_n} \cdot \cos\gamma = y \cdot v_m \cdot \sin(\alpha - \varphi - \beta) \cdot \cos(\gamma + \beta) \cdot \cos\gamma \qquad (14)$$

The velocity of D imposed by the joint is (15):

$$\dot{s}_D = v_D = \frac{v_m}{\cos\gamma \cdot l \cdot \cos\beta} \cdot [l \cdot \cos\beta \cdot \sin(\gamma + \alpha - \varphi) - a \cdot \cos\varphi \cdot \sin(\gamma + \beta)] \qquad (15)$$

3.3. The dynamics coefficient

The mechanism dynamic coefficient D is imposed to all gear and influences its function varying the crank rotation speed (the crank shaft rotation velocity). Any mechanism must take practical only one dynamic factor, D. To the engines in V the real dynamic coefficient is the result of a random momentary compromise between the two different dynamic coefficients imposed by the two pistons. For this reason the overall functioning of the V engine loud. The ideal solution (right) is obviously bringing the two dynamic factors to around or possibly even equal values. To this end were the two dynamic factors matched, to see what solutions exist to solve the obtained equation inα. The obtained expression is complex and has many variables (the various builder parameters of the engine in V). It sought an analytical synthesis using a complex computer program, by finding of the system alpha general solutions, regardless of the values of others constructive parameters, so that dynamic factors present equal values, and the engine so constructed to operate high efficiency without shocks and vibrations, without noise and with reduced noxious emissions, achieved with high power and lower fuel consumption. The cinematic chain composed of crankshaft, two pistons and two rods should function normally. The dynamic coefficient in D is (16):

$$\begin{cases} D_D = \dfrac{N}{n} \\ N = (1-x) \cdot l \cdot \sin(\alpha - \varphi - \beta) \cdot \cos(\gamma + \beta) \cdot \cos^2\gamma \cdot \cos\beta \\ n = l \cdot \cos\beta \cdot \sin(\gamma + \alpha - \varphi) - a \cdot \cos\varphi \cdot \sin(\gamma + \beta) \end{cases} \qquad (16)$$

It put the condition to have a single dynamic coefficient of the mechanism, D:

$$\begin{cases} D = D_D = D_B \Rightarrow x = \dfrac{N_x}{n_x} \\ N_x = l \cdot \sin(\alpha - \varphi - \beta) \cdot \cos(\gamma + \beta) \cdot \cos^2 \gamma \\ n_x = l \cdot \cos^2 \beta \cdot \sin(\gamma + \alpha - \varphi) - a \cdot \cos \beta \cdot \cos \varphi \cdot \sin(\gamma + \beta) + \\ l \cdot \sin(\alpha - \varphi - \beta) \cdot \cos(\gamma + \beta) \cdot \cos^2 \gamma \\ D = D_B = x \cdot \cos^2 \beta \end{cases} \quad (17)$$

The value of x was determined from the imposed condition to have a single dynamic coefficient for the mechanism.

4. DYNAMIC ANALYSIS

Analysis of dynamic system revealed a range of values for angle alpha that the theory exposed are likely to lead to the synthesis of V-optimal engine (see the table 1) [5].

Alfa angle values in grad **Table 1**

α [GRAD]
0 – 8
12 – 17
23 – 25
155 – 156
164 – 167
173 – 179

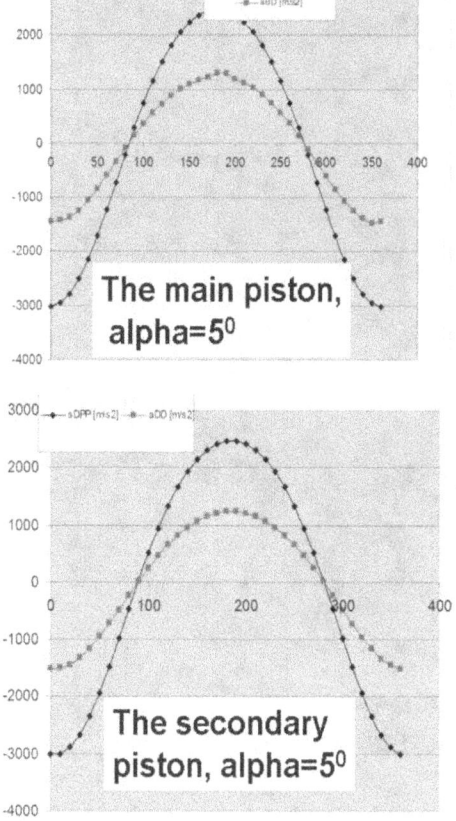
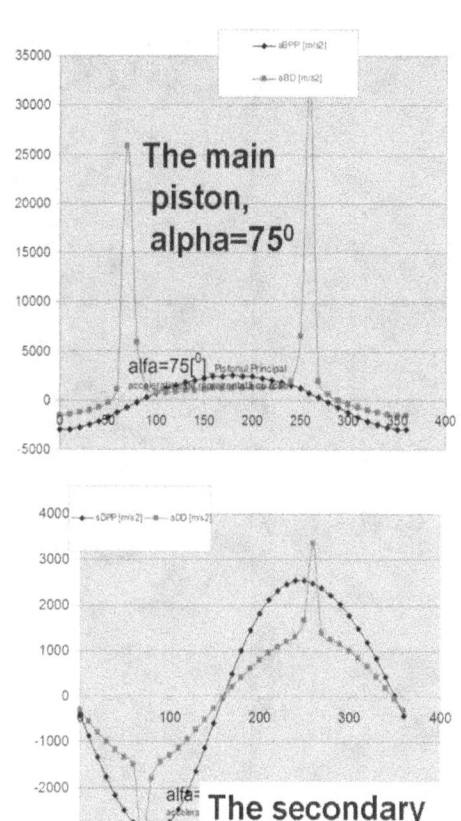

Fig. 2. Accelerations   **Fig. 3.** Accelerations

For some constructive parameters randomly taken (r=0.01 [m], l=0.1 [m], a=0.03 [m], b=0.07 [m]) and for a chosen speed of motor shaft (n=5000 [r/m]), it obtains three different diagrams for the displacement and acceleration of the pistons, corresponding to three alpha angles chosen randomly (5°, 75° and 95°), (see the figures 2-4).

The presented diagrams show only the two acceleration, $\ddot{s}$ (normal, cinematic acceleration, with blue) and a_D (special dynamic acceleration with pulses, with red), everyone for the main and second pistons.

Value of five degrees are at the beach of values indicated as appropriate, so that acceleration peaks hardly exceed the value of 1000 [m/s²] to both pistons (see the figure 2).

Diagrams in figures 3 and 4 are somewhat similar (but not identical) and present relevant cases also, even if the acceleration peaks have increased at about 3500 [m/s²] for the secondary piston and approximately 30,000 [m/s²] for the main piston.

The angles of 75 and 95 degrees can also be used (at least for the indicated constructive parameters), to take into account and ignition requirements uniformly distributed.

A V-engine which reaching local at the primary piston a peak of acceleration of 30000 [m/s²] to a motor shaft speed of 5000 [r/m] (it comes only a local impact) will work similar to engines in line but the power and efficiency higher.

However the use for alpha of constructive values shown in the table 1 may lead to the construction of a V engine quieter than the one in line.

The dynamic analysis made with the presented systems indicates some good values for the constructive angle (α), which allow the motor in V works normally without vibrations, noises and shocks (see the table 1):

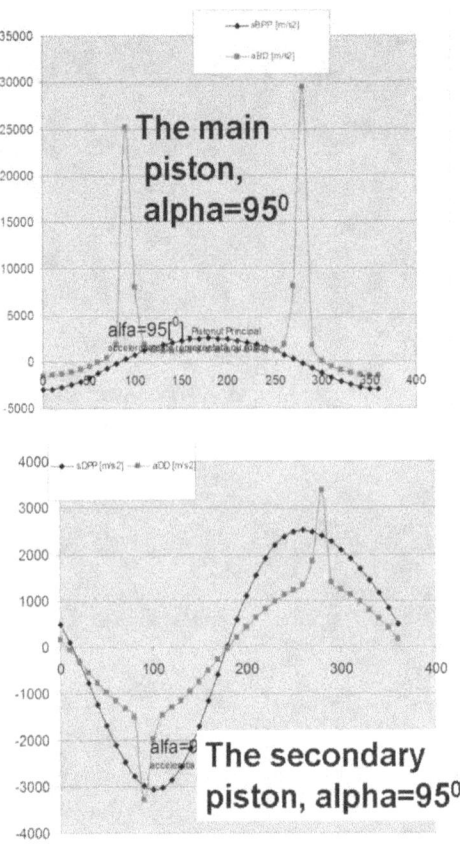

Fig. 4. *Accelerations*

5. CONCLUSIONS

SPECIFICATIONS: Acceleration diagrams presented were constructed based on an original method; they are the result of complex calculations with dynamic accelerations and dynamic coefficients which contain the vibrations and pulses; the relationships of calculation can not present classical accelerations known!

A-When shocks are very small, diagrams show even the accelerations.

b-When the shocks are visible the diagrams show the accelerations and their peaks.

c-When the shocks are large or very large, the diagrams will present only the shocks; in this case the accelerations overlapping shocks; accelerations are lower than the shocks and no longer can see (these cases but would not be desirable).

With the values in the table of constructive angle alpha can synthesize a quieter engine in V, regardless of the values of other constructive parameters of the engine.

A first observation arising from reading the values indicated for optimal alpha angle from the table, is that the values close to 90 degrees aren't present, and in general for these values design software looks a worse dynamics of engine in V. But that these values are used specifically to build engines in V, values determined by the number of cylinders and the condition for obtaining the ignitions uniformly distributed.

For the alpha values who do not appear in table, the built engine works with very large shocks which very difficult can be isolated even with the most modern rubber pads so that vibrations are felt in the vehicle interior, bringing with them uncomfortable and insecure amplified and by the unnatural noises produced by shock.

An important observation would be that today are used "new cinematic schemas of engines in V" (see the figure 5).

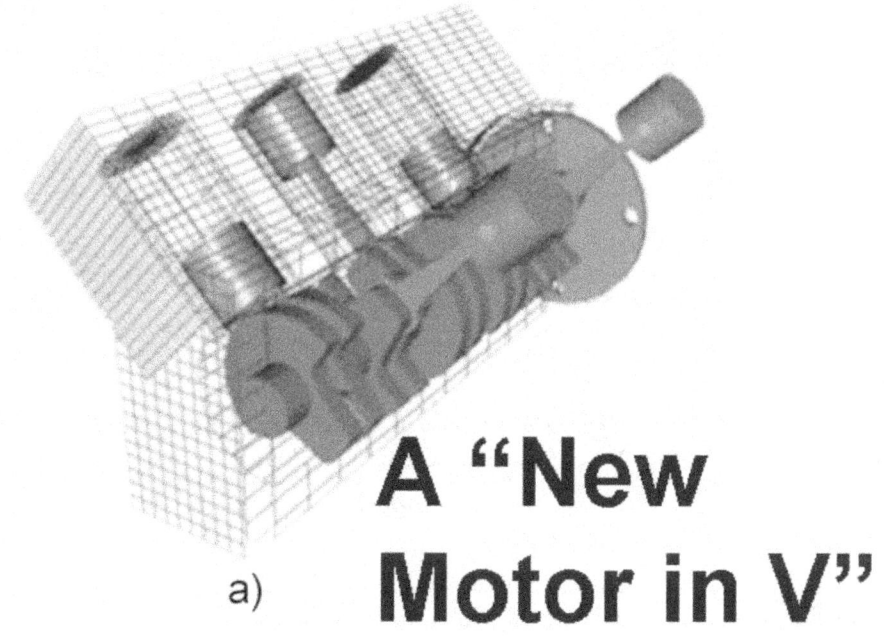

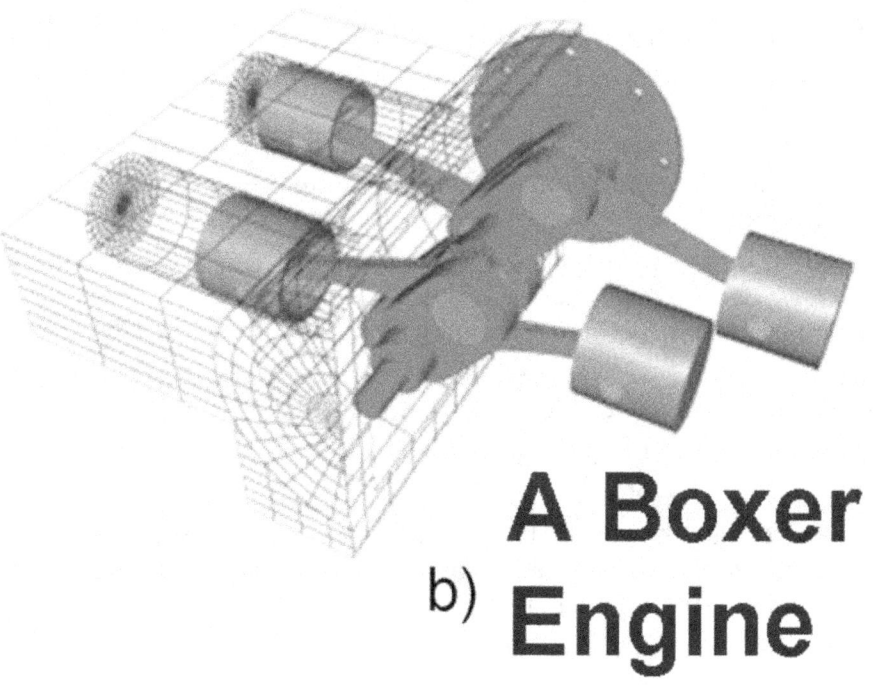

Fig. 5. *"New V Engines"*

These "new V Engines" to eliminate the vibrations have a single piston mounted on a crank throw and have inclined the axes first to right the second to left to give the appearance of the engine in V.

It's a pseudo-engine in V and the added efficiency disappears. The cylinders capacity should be increased to mimic the engine power in V, but also increases fuel consumption.

In this way, and the cylinders in line can be considered an engine in V with alpha of 0 degrees and boxer cylinders may be considered a V engine with alpha of 180 degrees.

With α indicate in the table 1 one can make V Engine work without vibrations.

REFERENCES

[1] GRUNWALD B., *Teoria, calculul și construcția motoarelor pentru autovehicule rutiere*. Editura didactică și pedagogică, București, 1980.

[2] Petrescu, F.I., Petrescu, R.V., *Câteva elemente privind îmbunătățirea designului mecanismului motor*, Proceedings of 8th National Symposium on GTD, Vol. I, p. 353-358, Brasov, 2003.

[3] Petrescu, F.I., Petrescu, R.V., *An original internal combustion engine*, Proceedings of 9th International Symposium SYROM, Vol. I, p. 135-140, Bucharest, 2005.

[4] Petrescu, F.I., Petrescu, R.V., *Determining the mechanical efficiency of Otto engine's mechanism*, Proceedings of International Symposium, SYROM 2005, Vol. I, p. 141-146, Bucharest, 2005.

[5] Petrescu, F.I., Petrescu, R.V., *V Engine Design*, Proceedings of International Conference on Engineering Graphics and Design, ICGD 2009, Cluj-Napoca, 2009.

[6]. FRĂȚILĂ, Gh., SOTIR, D., *PETRESCU, F., PETRESCU, V.*, ș.a. *Cercetări privind transmisibilitatea vibrațiilor motorului la cadrul și caroseria automobilului*. În a IV-a Conferință de Motoare, Automobile, Tractoare și Mașini Agricole, CONAT-matma, Brașov, 1982, Vol. I, p. 379-388.

[7]. MARINCAȘ, D., SOTIR, D., *PETRESCU, F., PETRESCU, V.*, ș.a. *Rezultate experimentale privind îmbunătățirea izolației fonice a cabinei autoutilitarei TV-14*. În a IV-a Conferință de Motoare, Automobile, Tractoare și Mașini Agricole, CONAT-matma, Brașov, 1982, Vol. I, p. 389-398.

[8]. FRĂȚILĂ, Gh., MARINCAȘ, D., BEJAN, N., FRĂȚILĂ, M., *PETRESCU, F., PETRESCU, R.*, RĂDULESCU, I. *Contributions a l'amelioration de la suspension du groupe moteur-transmission*. În buletinul Universității din Brașov, Seria A, Mecanică aplicată, Vol. XXVIII, 1986, p. 117-123.

[9]. Fjoseph L. Stout – Ford Motor Co., I. *Engine Excitation Decomposition Methods and V Engine Results*. In SAE 2001 Noise & Vibration Conference & Exposition, USA, 2001-01-1595, April 2001.

[10]. D. Taraza, "Accuracy Limits of IMEP Determination from Crankshaft Speed Measurements," *SAE Transactions, Journal of Engines* 111, 689-697, 2002.

[11]. FROELUND, K., S.G. FRITZ, and B. SMITH., *Ranking Lubricating Oil Consumption of Different Power Assemblies on an EMD 16-645E Locomotive Diesel Engine*. Presented at and published in the Proceedings of the 2004 CIMAC Conference, Kyoto, Japan, June 2004.

[12]. Leet, J.A., S. Simescu, K. Froelund, L.G. Dodge, and C.E. Roberts Jr., *Emissions Solutions for 2007 and 2010 Heavy-Duty Diesel Engines*. Presented at the SAE World Congress and Exhibition, Detroit, Michigan, March 2004. SAE Paper No. 2004-01-0124, 2004.

www.ingramcontent.com/pod-product-compliance
Lightning Source LLC
Chambersburg PA
CBHW081815220526
45470CB00007B/2323